ISBN 978-0-265-64193-4
PIBN 10249191

LIFE AND DEATH

BY

A. DASTRE,

PROFESSOR OF PHYSIOLOGY AT THE SORBONNE.

TRANSLATED BY

W. J. GREENSTREET, M.A., F.R.A.S.

THE WALTER SCOTT PUBLISHING CO., LTD.,

PATERNOSTER SQUARE, LONDON, E.C.

CHARLES SCRIBNER'S SONS,

153-157 FIFTH AVENUE, NEW YORK.

1911

PREFACE.

THE educated and inquiring public of the present day addresses to the experts who have specialized in every imaginable subject the question that was asked in olden times of Euclid by King Ptolemy Philadelphus, Protector of Letters. Recoiling in dismay from the difficulties presented by the study of mathematics and annoyed at his slow progress, he inquired of the celebrated geometer if there was not some royal road, could he not learn geometry more easily than by studying the Elements. The learned Greek replied, "There is no royal road." These royal roads making every branch of science accessible to the cultivated mind did not exist in the days of Ptolemy and Euclid. But they do exist to-day. These roads form what we call Scientific Philosophy.

Scientific philosophy opens a path through the hitherto inextricable medley of natural phenomena. It throws light on facts, it lays bare principles, it replaces contingent details by essential facts. And thus it makes science accessible and communicable. Intellectually it performs a very lofty function.

There is virtually a philosophy of every science.

There is therefore a philosophy of the science which deals with the phenomena of life and death—*i.e.*, of physiology. I have endeavoured to give a summary of this philosophy in this volume. I have had in view two classes of readers. In the first place there are readers of general culture who are desirous of knowing something of the trend of ideas in biology. They already form quite a large section of the great public.

These scholars and inquirers, with Bacon, believe that the only science is general science. What they want to know is not what instruments we use, our processes, our technique, and the thousand and one details of the experiments on which we spend our lives in the laboratory. What they are interested in are the general truths we have acquired, the problems we are trying to solve, the principles of our methods, the progress of our science in the past, its state in the present, its probable course in the future.

But I venture to think that this book is also addressed to another class of readers, to those whose professional study is physiology. To them it is dedicated. They have been initiated into the mysteries of the science. They are learning it by practice. That is the right method. Practice makes perfect. Claude Bernard used to say that in order to be an expert in experimental science you must first be "a laboratory rat." And among us there are many such "laboratory rats." They are guided in the daily task of investigation by a dim instinct of the path and of the direction of contemporary physiology. Perhaps it may be of assistance to them to find their more or less unconscious ideas here expressed in an explicit form.

A. DASTRE.

CONTENTS.

BOOK I.

The Frontiers of Science. General Theories of Life and Death. Their Successive Transformations.

BOOK II.

The Doctrine of Energy and the Living World. General Ideas of Life. Alimentary Life.

BOOK III.

The Characters Common to Living Beings.

BOOK IV.

THE LIFE OF MATTER.

BOOK V.

SENESCENCE AND DEATH.

LIFE AND DEATH.

BOOK I.

THE FRONTIERS OF SCIENCE—GENERAL THEORIES OF LIFE AND DEATH—THEIR SUCCESSIVE TRANSFORMATIONS.

CHAPTER I.

EARLY THEORIES.

Animism — Vitalism — The Physico-Chemical Theory — Their Survival and Transformations.

THE fundamental theories of science are but the expression of its most general results. What, then, is the most general result of the development of physiology or biology—that is to say, of that department of science which has life as its object? What glimpse do we get of the fruit of all our efforts? The answer is evidently the response to that essential question—What is Life?

There are beings which we call living beings; there are bodies which have never been alive—inanimate bodies; and there are bodies which are no longer alive—dead bodies. The fact that we use these terms implies the idea of a common attribute, of a *quid proprium*, life, which exists in the first, has never existed in the second, and has ceased to exist in the last. Is this idea correct? Suppose for a moment that this is so, that this implicit supposition has a foundation, and that there really is something which corresponds to the word "*life*." Must we then wait for the last days of physiology, and in a measure for its last word before we know what is hidden behind this word, "life"?

Yes, no doubt positive science should be precluded from dealing with questions of this kind, which are far too general. It should be limited to the study of second causes. But, as a matter of fact, scientific men in no age have entirely conformed to this provisional or definitive antagonism. As the human mind cannot rest satisfied with indefinite attempts, or with ignorance pure and simple, it has always asked, and even now asks, from the spirit of system the solution which science refuses. It appeals to philosophical speculation. Now, philosophy, in order to explain life and death, offers us hypotheses. It offers us the hypotheses of thirty, of a hundred, or two thousand years ago. It offers us animism; vitalism in its two forms, unitary vitalism or the doctrine of vital force, and dismembered vitalism or the doctrine of vital properties; and finally, materialism, a mechanical theory, unicism or monism,—to give it all its names—*i.e.*, the physico-chemical doctrine of life. There are, therefore, at the present day, in biology, representatives of these three systems which have never agreed on the explanation of vital phenomena—namely, animists, vitalists, and monists.

But it is pretty clear that there must have been some change between yesterday and to-day. Not in vain has general science and biology itself made the progress which we know has been made since the Renaissance, and especially during the course of the nineteenth century. The old theories have been compelled to take new shape, such parts as have become obsolete have been cut away, another language is spoken—in a word, the theories have become rejuvenated. The neo-animists of our day, Chauffard in 1878, von Bunge in 1889, and more recently Rindfleisch, do not hold exactly the same views as Aristotle, St. Thomas Aquinas, or Stahl. Contemporary neo-vitalists, physiologists like Heidenhain, chemists like Armand Gautier, or botanists like Reinke do not between 1880 and 1900 hold the same views as Paracelsus in the fifteenth century and Van Helmont in the seventeenth, as Barthez and Bordeu at the end of the eighteenth, or as Cuvier and Bichat at the beginning of the nineteenth century. Finally, the mechanicians themselves, whether they be disciples of Darwin and Haeckel, as most biologists of our own time, or disciples of Lavoisier, as most physiologists of the present day, have passed far beyond the ideas of Descartes. They would reject the coarse materialism of the celebrated philosopher. They would no longer consider the living organism as a machine, composed of nothing but wheels, springs, levers, presses, sieves, pipes, and valves; or again of matrasses, retorts, or alembics, as the iatro-mechanicians and would-be chemists of other days believed.

All that is changed, at any rate in form. If we look back only thirty or forty years we see that the old doctrines have undergone more or less profound modifications. The changes of form, which have been made necessary by the acquisitions of contemporary science, enable us to appreciate its progress. They enable us to give an account of the progress of biology, and for this reason they deserve to be examined with some attention. It is into this examination that I ask my readers to accompany me.

CHAPTER II.

ANIMISM.

The Common Characteristic of Animism and Vitalism: the Human Statue—Primitive Animism—Stahl's Animism—First Objection with Reference to the Relation between Soul and Body—Second Objection: the Unconscious Character of Vital Operations—Twofold Modality of the Soul—Continuity of the Soul and Life.

CHILDREN are taught that there are three kingdoms in Nature—the mineral kingdom and the two living kingdoms, animal and vegetable. This is the whole of the sensible world. Then above all that is placed the world of the soul. School-boys therefore have no doubts on the doctrines that we discuss here. They have the solution. To them there are three distinct spheres, three separate worlds—matter, life, and thought.

It is this preconceived idea that we are about to examine. Current opinion solves *a priori* the question of the fundamental homogeneity or lack of resemblance of these three orders of phenomena—the phenomena of inanimate nature, of living nature, and of the thinking soul. *Animism*, *vitalism*, and *monism* are, in reality, different ways of looking at them. They are the different answers to this question:—Are vital, psychic, and physico-chemical manifestations essentially distinct? Vitalists distinguish between life and thought, animists identify them. In the opposite camp mechanicians, materialists, or monists make the same mistake as the animists, but to that mistake they add another: they assimilate the forces at play in animals and plants to the general forces of the universe; they confuse all three—soul, life, inanimate nature.

These problems belong on many sides to metaphysical speculation. They have been discussed by philosophers; they have been solved from time immemorial in different ways, for reasons and by arguments which it is not our purpose to examine here, and which, moreover, have not changed. But on some sides they belong to science, and must be tested in the light of its progress. Cuvier and Bichat, for example, considered that the forces in action in living beings were not only different from physico-mechanical forces, but were utterly opposed to them. We now know that this antagonism does not exist.

The preceding doctrines, therefore, depend up to a certain point on experiment and observation. They are subject to the test of experiment and observation in proportion as the latter can give us information on the degree of difference or analogy presented by psychic, vital, and physico-chemical facts. Now, scientific investigations have thrown light on these points. There is no doubt that the analogies and the resemblances of these three orders of manifestations have appeared more and more numerous and striking as our knowledge has advanced. Hence it is that animism can count to-day but very few advocates in biological science. Vitalism in its different forms counts more supporters, but the great majority have adopted the physico-chemical theory.

Both animism and vitalism separate from matter a directing principle which guides it. At bottom they are mythological theories somewhat similar to the paganism of old. The fable of Prometheus or the story of Pygmalion contains all that is essential. An immaterial principle, divine, stolen by the Titan from Jupiter, or obtained from Venus by the Cypriot sculptor, descends from Olympus and animates the form, till then inert, which has been carved in the marble or modelled in the clay. In a word, there is a human statue. It receives a breath of heavenly fire, a vital force, a divine spark, a soul, and behold! it is alive. But this breath can also leave it. An accident happens, a clot in a vein, a grain of lead in the brain—the life escapes, and all that is left is a corpse. A single instant has proved sufficient to destroy its fascination. This is how all men picture to their minds the scene of death. The breath escapes; something flies away, or flows away with the blood. The happy genius of the Greeks conceived a graceful image of this, for they represented the life or the soul in the form of a butterfly (Psyche) leaving the body, an ethereal butterfly, as it were, opening its sapphire wings.

But what is this subtle and transient guest of the human statue, this passing stranger which makes of the living body an inhabited house? According to the animists it is the soul itself, in the sense in which the word is understood by philosophers; the immortal and reasoning soul. To the vitalists it is an inferior, subordinate soul; a soul, as it were, of secondary majesty, the vital force, or in a word, life.

Primitive Animism.—Animism is the oldest and most primitive of the conceptions presented to the human mind. But in so far as it is a co-ordinated doctrine, it is the most recent. In fact it only received its definitive expression in the eighteenth century, from Stahl, the philosopher-physician and chemist.

According to Tylor, one of the first speculations of primitive man, of the savage, is as to the difference between the living body and the corpse. The former is an inhabited house, the latter is empty. To such rudimentary intellects the mysterious inhabitant is a kind of *double* or duplicate of the human form. It is only revealed by the shadow which follows the body when illuminated by the sun, by the image of its reflection in the water, by the echo which repeats the voice. It is only seen in a dream, and the figures which people and animate our dreams are nothing but these doubled, impalpable beings. Some savages believe that at the moment of death the double, or the soul, takes up its residence in another body. Sometimes each individual possesses, not one of these souls, but several. According to Maspero, the Egyptians counted at least five, of which the principle, the *ka* or *double*, would be the aeriform or vaporous image of the living form. Space is peopled by souls on their travels, which leave one set of bodies to occupy another set. After having been the cause of life in the bodies which they animated, they react from without on other beings, and are the cause of all sorts of unexpected events. They are benevolent or malevolent spirits.

Analogy inevitably leads simple minds to extend the same ideas to animals and plants; in a word, to attribute souls to everything alive, souls more or less nomadic, wandering, or interchangeable, as is taught in the doctrine of metempsychosis. Mons. L. Errera points out that this primitive, co-ordinated, hierarchized doctrine—meet subject for the poet's art—is the basis of all ancient mythologies.

The Animism of Stahl.—Modern animism was much more narrow in scope. It was a medical theory—*i.e.* almost exclusive to man. Stahl had adopted it in a kind of reaction against the exaggerations of the mechanical school of his time. According to him, the life of the body is due to the intelligent and reasoning soul. It governs the corporeal substance and directs it towards an assigned end. The organs are its instruments. It acts on them directly, without intermediaries. It makes the heart beat, the muscles contract, the glands secrete, and all the organs perform their functions. Nay more, it is itself the architectonic soul, which has constructed and which maintains the body which it rules. It is the *mens agitat molem* of Virgil.

It is remarkable that these ideas, so excessively and exaggeratedly spiritualistic, should have been brought forward by a chemist and a physician, while ideas completely opposed to these were admitted by philosophers like Descartes and Leibniz, who were decided believers in the spirituality of the soul. Stahl had been Professor of Medicine at the University of Halle, physician to the Duke of Saxe-Weimar, and later to the King of Prussia. He left an important medical and chemical work, both theoretical and practical. He is the author of the celebrated theory of phlogiston, which held its ground in chemistry up to the time of Lavoisier. He died about 1734.

Animism survived him for some time, maintained by the zeal of a few faithful disciples. But after the witty mockery of Bordeu,[1] in 1742, it began to decay. We must, however, point out that an attempt to revive this theory was made in 1878 by a well-known doctor of the last generation, E. Chauffard. While preserving the essential features of the theory, this learned physician proposed to bring it into harmony with modern science, and to free it from all the reproaches which had been levelled at it.

The Animism of E. Chauffard.—These reproaches were numerous. The most serious is of a philosophic nature. It rises from the difficulty of conceiving a direct and immediate action of the soul, considered as a spiritual principle, upon the matter of the body. There is such an abyss—hewn by the philosophic mind itself—between soul and body, that it is impossible to imagine any relation between them. We can only get a glimpse of how the soul might become an instrument of action.

This was the problem which sorely tried the genius of Leibniz. Descartes, in earlier days, attacked it vigorously, like an Alexander cutting the Gordian knot. He separated the soul from the body, and made of the latter a pure machine in the government of which the soul had no part. He attributed all the known manifestations of vital activity to inanimate forces. Leibniz, also, was compelled to reject all action, all contact, all direct relation, every real bond between soul and body, and to imagine between them

[1] In a thesis presented in 1742 at Montpellier, Bordeu, then only twenty years of age, made game of the tasks imposed by animists on the Soul, "which has to moisten the lips when required;" or, "whose anger produces the symptoms of certain diseases;" or again, "which is prevented by the consequences of original sin from guiding and directing the body."

a purely metaphysical relation—pre-established harmony:—"Soul and body agree in virtue of this harmony, the harmony pre-established since the creation, and in no way by a mutual, actual, physical influence. Everything that takes place in the soul takes place as if there were no body, and so everything takes place in the body as if there were no soul." At this point we almost reach a scientific materialism. It is easy for the materialist to break this frail tie of pre-established harmony which so loosely unites body and soul, and to exhibit the organism as under the sole control of universal mechanics and physics.

Thus the weak point of Stahl's animism was the supposition of a direct action exercised on the organism by a distinct, heterogeneous, spiritual principle.

Chauffard has endeavoured to avoid this pitfall. In conformity with modern ideas, he has brought together what the ancient philosophers and Stahl himself separated—the activity of matter and the activity of the soul. "Thought, action, function, are embraced in an indissoluble union." This is the classical but not very lucid theory which has been so often reproduced—*Homo factus est anima vivens*—which Bossuet has expressed in the celebrated formula: "Soul and body form a natural whole."

A second objection raised against animism is that the soul acts consciously, with reflection, and with volition, and that its essential attributes are not found in most physiological phenomena, which, on the contrary are automatic, involuntary, and unconscious. The contradictory nature of these characteristics has obliged vitalists to conceive of a vital principle distinct from thought. Chauffard, agreeing here with

Boulliez, Tissot, and Stahl himself, does not accept this distinction; he refuses to shatter the unity of the vivifying and thinking principle. He prefers to attribute to the soul two modes of action: the one which is exercised on the acts of thought, and hence it proceeds consciously, with reflection, and with volition; the other exercising control over the physiological phenomena which it governs, "by unconscious impressions, and by instinctive determinations, obeying primordial laws." This soul is hardly in keeping with his definition of a conscious, reflecting, and voluntary principle; it is a new soul, a somatic soul, singularly akin to that *rachidian soul* which, according to Pflüger, a well-known German physiologist, resides in each segment of the spinal marrow, and is responsible for reflex movements.

Twofold Modality of the Soul.—This twofold modality of the soul, this duality admitted by Stahl and his disciples, was repugnant to many thinkers, and it is this repugnance that gave rise to the vitalistic school. It appeared to them to be a heresy tainted by materialism—and so it was. In this lay the strength and the weakness of animism. It admits of a unique animating principle for all the manifestations of the living being, for the higher facts in the realm of thought, and for the lower facts connected with the body. It throws down the barriers which separate them. It fills up the gap between the different forms of human activity, and assimilates them the one to the other.

Now this is precisely what materialism does. It, too, reduces to a single order the psychical and physiological phenomena, between which it no longer recognizes anything but a difference of degree, thought being only a maximum of the vital movement, or life a minimum of thought. In truth, the aims of the two schools are diametrically opposed; the one claims to raise corporeal activity to the dignity of thinking activity, and to spiritualize the vital fact; the other lowers the former to the level of the latter and materializes the psychic fact. But, though the intentions are different, the result is identical. Spiritualistic monism inclines towards materialistic monism. One step more, and the soul, confused with life, will be confused with physical forces.

On the other hand, twofold modality has this advantage, that it escapes the objection drawn from the existence of so many living beings to which a thinking soul cannot be attributed; an anencephalous fœtus, the young of the higher animals, the lower animals and plants, living without thought, or with a minimum of real, conscious thought. The advocate of animism replies that this physiological activity is still a soul, but one which is barely aware of its existence—a gleam of consciousness. In this theory, the knowledge of self, the consciousness, is of all degrees. On the other hand, in the eyes of the vitalist, it is an absolute fact which allows of no attenuation, of no middle course between the being and the non-being.

It is this conception of the continuity of the soul and life, it is the affirmation of a possible lowering of the complete consciousness down to a mere gleam of knowledge, and finally down to unconscious vital activity, which saved animism from complete shipwreck. That is why this ancient doctrine finds, even in the present day, a few rare supporters. An able German scientist,

G. von Bunge, well known for his researches in physiological chemistry, professes animistic views in a work which appeared in 1889. He attributes to organized beings a guiding principle, a kind of vital soul. A distinguished naturalist, Rindfleisch, of Lübeck, has likewise taken his place among the advocates of what we may call neo-animism.

CHAPTER III.

VITALISM.

Its Extreme Forms—Early Vitalism, and Modern Neo-vitalism—Advantage of distinguishing between Soul and Life—§ 1. *The Vitalism of Barthez*—Its Extension—The Seat of the Vital Principle—The Vital Knot—The Vital Tripod—Decentralisation of the Vital Principle—§ 2. *The Doctrine of Vital Properties*—Galen, Van Helmont, Xavier Bichat, and Cuvier—Vital and Physical Properties antagonistic—§ 3. *Scientific Neo-vitalism*—Heidenhain—§ 4. *Philosophical Neo-vitalism*—Reinke.

Extreme Forms: Early Vitalism and Modern Neo-vitalism.—Contemporary neo-vitalism has weakened primitive vitalism in some important points. The latter made of the vital fact something quite specific, irreducible either to the phenomena of general physics or to those of thought. It absolutely isolated life, separating it above from the soul, and below from inanimate matter. This sequestration is nowadays much less rigorous. On the psychical side the barrier remains, but it is lowered on the material side. The neo-vitalists of to-day recognize that the laws of physics and chemistry are observed within, as well as without, the living body; the same natural forces intervene in both, only they are "otherwise directed."

The vital principle of early times was a kind of anthromorphic, pagan divinity. To Aristotle, this force, the *anima*, *the Psyche*, worked, so to speak, with human hands. According to the well-known expression, its situation in the human body corresponds to that of a pilot on a vessel, or to that of a sculptor or his assistant before the marble or clay. And, in fact, we have no other clear image of a cause external to the object. We have no other representation of a force external to matter than that which is offered by the craftsman making an object, or in general by the human being with his activity, free, or supposed to be free, and directed towards an end to be realized.

Personifications of this kind, the mythological entities, the imaginary beings, the ontological fictions, which ever filled the stage in the mind of our predecessors, have definitely disappeared; no longer have they a place in the scientific explanations of our time. The neo-vitalists replace them by *the idea of direction*, which is another form of the same idea of finality. The series of second causes in the living being seems to be regulated in conformity with a plan, and directed with a view to carrying it out. The tendency which exists in every being to carry out this plan,—that is to say, the tendency towards its end,—gives the impulse that is necessary to carry it out. Neo-vitalists claim that vital force directs the phenomena which it does not produce, and which are in reality carried out by the general forces of physics and chemistry.

Thus, the directing impulse, *considered as really active*, is the last concession of modern vitalism. If we go further, and if we refuse to the directing idea executive power and efficient activity, the vital principle is weakened, and we abandon the doctrine. We can no longer invoke it. We cease to be vitalists if the part played by the vital principle is thus far restricted. At first it was both the author of the plan and the universal architect of the organic edifice; it is now only the architect directing his workmen, and they are physical and chemical agents. It is now reduced to the plan of the work, and even this plan has no objective existence; it is now only an *idea.* It has only a shadow of reality. To this it has been reduced by certain biologists. For this we may thank Claude Bernard; and he has thereby placed himself outside and beyond the weakest form of vitalism. He did not consider the *idea of direction* as a real principle. The connection of phenomena, their harmony, their conformity to a plan grasped by the intellect, their fitness for a purpose known to the intellect, are to him but a mental necessity, a metaphysical concept. The plan which is carried out has only a subjective existence; the directing force has no efficient virtue, no executive power; it does not emerge from the intellectual domain in which it took its rise, and does not "react on the phenomena which enabled the mind to create it."

It is between these two extreme incarnations of the vital principle, on the one hand an executive agent, on the other a simple directing plan, that the motley procession of vitalist doctrines passes on its way. At the point of departure we have a vital force, personified, acting, as we have stated, as if with human hands fashioning obedient matter; this is the pure and primitive form of the theory. At the other extreme we have a vital force which is now only a directing idea, without objective existence, and without an executive rôle; a mere concept by which the mind gathers together and conceives of a succession of physico-chemical phenomena. On this side we are brought into touch with monism.

The Reasons given by the Vitalists for distinguishing Soul from Life.—It is, in particular, on the opposite side, in the psychical world, that the early vitalists professed to entrench themselves. We have just seen that their doctrines were not so subtle as those of to-day; the vital principle to them was a real agent, and not an ideal plan in the process of being carried out. But they distinguished this spiritual principle from another co-existent with it in superior living beings—at any rate, in man: the thinking soul. They boldly distinguished between them, because the activity of the one is manifested by knowledge and volition, while on the contrary, the manifestations of the other for the most part escape both consciousness and volition.

In fact, we know nothing of what goes on in the normal state of our organs. Their perfect performance of their functions is translated to us solely by an obscure feeling of comfort. We do not feel the beating of the heart, the periodic dilations of the arteries, the movements of the lungs or intestine, the glands at their work of secretion, or the thousand reflex manifestations of our nervous system. The soul, which is conscious of itself, is nevertheless ignorant of all this vital movement, and is therefore external to it.

This is the view of all the philosophers of antiquity. Pythagoras distinguished the real soul, the thinking soul, the *Nous*, the intelligent and immortal principle, characterized by the attributes of consciousness and volition, from the vital principle, the *Psyche*, which gives breath and animation to the body, and which is a soul of secondary majesty, active, transient, and mortal. Aristotle did the same. On the one side he placed the soul properly so called, the *Nous* or intellect—that is to say, the understanding with its rational intelligence; on the other side was the directing principle of life, the irrational and vegetative Psyche.

This distinction agrees with the fact of the diffusion of life. Life does not belong to the superior animals alone, and to the man in whom we can recognize a reasoning soul. It is extended to the vast multitude of humbler beings to which such lofty faculties cannot be attributed, the invertebrates, microscopic animals, and plants. The advantage is compensated for by the inconvenience of breaking down all continuity between the soul and life; a continuity which is the principle of the two other doctrines, animism and monism, and which is, we may say, the very aim and the unquestionable tendency of science.

As for classical philosophy, it satisfies the necessity of establishing the unity of the living being,—*i.e.*, of bringing into harmony soul and body,—but in a manner which we need not here discuss. It attributes to the soul several modalities, several distinct powers: powers of the vegetative life, powers of the sensitive life, and powers of the intellectual life. And this other solution of the problem would be, in the opinion of M. Gardair, in complete agreement with the doctrines of St. Thomas Aquinas.

§ I. THE VITALISM OF BARTHEZ: ITS EXTENSION.

Vitalism reached its most perfect expression in the second half of the eighteenth century in the hands of the representatives of the Montpellier school—Bordeu, Grimaud, and Barthez. The last, in particular, contributed to the prevalence of the doctrine in medical circles. A man of profound erudition, a collaborateur with d'Alembert in the *Encyclopædia*, he exercised quite a preponderant influence on the medicine of his day. Stationed at Paris during part of his career, physician to the King and the Duke of Orleans, we may say that he supported his theories by every imaginable influence which might contribute to their success. In consequence of this, the medical schools taught that vital phenomena are the immediate effects of a force which has no analogues outside the living body. This conception reigned unchallenged up to the days of Bichat.

After Bichat, the vitalism of Barthez, more or less modified by the ideas of the celebrated anatomist, continued to hold its own in all the schools of Europe until about the middle of the nineteenth century. Johannes Müller, the founder of physiology in Germany, admitted, about 1833, the existence of a unique vital force "aware of all the secrets of the forces of physics and chemistry, but continually in conflict with them, as the supreme cause and regulator of all phenomena." When death came, this principle disappeared and left no trace behind. One of the founders of biological chemistry, Justus Liebig, who died in 1873, shared these ideas. The celebrated botanist, Candolle, who lived up to 1893, taught at the beginning of his career that the vital force was one of the four forces ruling in nature, the other three being—attraction, affinity, and intellectual force. Flourens, in France, made the vital principle one of the five properties of forces residing in the nervous system. Another contemporary, Dressel, in 1883, endeavoured to bring back into fashion this rather primitive, monistic, and efficient vitalism.

The Seat of the Vital Principle.—Meanwhile, another question was asked with reference to this vital principle. It was a question of ascertaining its seat: or, in other words, of finding its place in the organism. Is it spread throughout the organism, or is it situated in some particular spot from which it acts upon every part of the body? Van Helmont, a celebrated scientist at the end of the sixteenth century, who was both physician and alchemist, gave the first and rather quaint solution of this difficulty. The vital principle, according to him, was situated in the stomach, or rather in the opening of the pylorus. It was the *concierge*, so to speak, of the stomach. The Hebrew idea was more reasonable. The life was connected with the blood, and was circulated with it by means of all the veins of the organism. It escaped from a wound at the same time as the liquid blood. It is clear that in this belief we see why the Jews were forbidden to eat meat which had not been bled.

The Vital Knot.—In 1748 a doctor named Lorry found that a very small wound in a certain region of the spinal marrow brought on sudden death. The position of this remarkable point was ascertained in 1812 by Legallois, and more accurately still by Flourens in 1827. It is situated in the rachidian bulb, at the level of the junction of the neck and the head; or more precisely, on the floor of the fourth ventricle, near the origin of the eighth pair of cranial nerves. This is what was called the *vital knot.* Upon the integrity of this spot, which is no bigger than the head of a pin, depends the life of the animal. Those who believed in a localisation of the vital principle thought that they had found the seat desired; but for that to be so the destruction of this spot must be irremediable, and must necessarily cause death.. But if the *vital knot* be destroyed, and respiration be artificially induced by means of a bellows, the animal resists: it continues to live. It is only the nervous stimulating mechanism of the respiratory movements which has been attacked in one of its essential parts.

Life, therefore, resides no more in this point than it does in the blood or in the stomach. Later experiment has shown that it resides everywhere, that each organ enjoys an independent life. Each part of the body is, to use Bordeu's strong expression, "*an animal in an animal*"; or to adopt the phrase due to Bichat, "*a particular machine within the general machine.*"

The Vital Tripod.—What then is life, or, in other words, what is the biological activity of the individual, of the animal, of man? It is clearly the sum total, or rather, the harmony of these partial lives of the different organs. But in this harmony it seems that there are certain instruments which dominate and sustain the others. There are some whose integrity is more necessary to the preservation of existence and health, and of which any lesion makes death more inevitable. They are the lungs, the heart, and the brain. Death always ensues, said the early doctors, if any one of these three organs be injured. Life depends, therefore, on them, as if upon a three-legged support. Hence the idea of the *vital tripod.* It is no longer a single seat for the vital principle, but a kind of throne on three-supports. Life is decentralized.

This was only the first step, very soon followed by many others, in the direction of vital decentralization. Experiment showed, in fact, that every organ separated from the body will continue to live if provided with the proper conditions. And here, it is not only a question of inferior beings; of plants that are propagated by slips; of the *hydra* which Trembley cut into pieces, each of which generated a complete hydra; of the *naïs* which C. Bonnet cut up into sections, each of which reconstituted a complete annelid. There is no exception to the rule.

Decentralization of the Vital Principle.—The result is the same in the higher vertebrates, only the experiment is much more difficult. At the Physiological Congress of Turin in 1901, Locke showed the heart of a hare, extracted from the body of the animal, and beating for hours as energetically and as regularly as if it were in its place. He suspended it in the air of a room at the normal temperature, the sole condition being that it was irrigated with a liquid composed of certain constituents. The animal had been dead some time. More recently Kuliabko has shown in the same way the heart of a man still beating, although the man had been dead some eighteen hours. The same experiment is repeated in any physiological laboratory, in a much easier manner, with the heart of a tortoise. This organ, extracted from the body, fitted up with rubber tubes to represent its arteries and veins, and filled with the defibrinated blood of a horse or an ox taken from the slaughter-house, works for hours and days pumping the liquid blood into its rubber aorta, just as if it were pumping it into the living aorta.

But it is unnecessary to multiply examples. Every organ can be made to live for a longer or shorter period even though removed from its natural position; muscles, nerves, glands, and even the brain itself. Each organ, each tissue therefore enjoys an independent existence; it lives and works for itself. No doubt it shares in the activity of the whole, but it may be separated therefrom without being thereby placed in the category of dead substances. For each aliquot part of the organism there is a partial life and a partial death.

This decentralization of the vital activity is finally extended in complex beings from the organs to the tissues, and from the tissues to the anatomical elements—the cells. The idea of decentralization has given birth to the second form of vitalism, a softened down and weakened form—namely, pluri-vitalism, or the theory of vital properties.

§ 2. The Theory of Vital Properties.

The advocates of the theory of vital properties have cut up into fragments the monistic and indivisible guiding principle of Bordeu and Barthez. They have given it new currency—pluri-vitalism. This theory maintains the existence of spiritual powers of a lower order, which control phenomena more intimately than the vital principle did. These powers, less lofty in their dignity than the rational soul of the animists, or the soul of secondary majesty of the unitarian vitalists, are eventually incorporated in the living matter of which they will then be no longer more than the properties. Brought into closer connection therefore with the sensible world, they will be more in harmony with the spirit of research and with scientific progress.

The defect of the earlier conceptions, their common illusion, rose from their seeking the cause outside the object, from their demanding an explanation of vital phenomena from a principle external to living, immaterial, and unsubstantial matter. Here this defect is less marked. The pluri-vitalists will in turn appeal to the vital properties as modes of activity, inherent in the living substance in which and by which they are manifested, and derived from the arrangement of the molecules of this substance—that is to say, from its organization. This is almost the conception of the present day.

But this progress will only be realized at the end of the evolution of the pluri-vitalist theory. At the outset this theory seems an exaggeration of its predecessor, and a still more exaggerated form of the mythological paganism with which it was reproached. The archeus, the blas, the properties, the spirits—all have at first the effect of the genii or of the gods imagined by the ancients to preside over natural phenomena, of Neptune stirring up the waters of the sea, and of Eolus unchaining the winds. These divinities of the ancient world, the nymphs, the dryads, and the sylvan gods, seem to be transported to the Middle Ages, to that age of argument, that philosophical period of the history of humanity, and there metamorphosed into occult causes, immaterial powers, and personified forces.

Galen.—The first of the pluri-vitalists was Galen, the physician of Marcus Aurelius, the celebrated author of an Encyclopædia of which the greater part has been lost, and of which the one book preserved held its own as the anatomical oracle and breviary throughout the Middle Ages. According to Galen the human machine is guided by three kinds of spirits: *animal spirits*, presiding over the activity of the nervous system; *vital spirits* governing most of the other functions; and finally, *natural spirits* regulating the liver and susceptible of incorporation in the blood. In the sixteenth century, in the time of Paracelsus, Galen's spirits became *Olympic spirits.* They still presided over the functional activity of the organs, the liver, heart, and brain, but they also existed in all the bodies of nature.

Van Helmont.—Finally, the theory was laid down by Van Helmont, physician, chemist, experimentalist, and philosopher, endowed with a rare and penetrating intellect. Here we find many profound truths combined with fantastic dreams. Refusing to admit the direct action of an immaterial agent, such as the soul, on inert matter, on the body, he filled up the abyss which separated them by creating a whole hierarchy of immaterial principles which played the part of mediators and executive agents. At the head of this hierarchy was placed the thinking and immortal soul; below was the sensitive and mortal soul, having for its minister the *principal archeus*, the *aura vitalis*, a kind of incorporeal agent, which is remarkably like the vital principle, and which had its seat at the orifice of the stomach. Below again were the subordinate agents, the *blas*, or *vulcans* placed in each organ, and intelligently directing its mechanism like skilful workmen.

These chimerical ideas are not, however, so far astray as the theory of vital properties. When we see a muscle contract, we say that this phenomenon is due to a vital property—*i.e.*, a property without any analogue in the physical world, namely *contractility*, In the same way the nerve possesses two vital properties, *excitability* and *conductibility*, which Vulpian proposed to blend into one, calling it *neurility*. These are mere names, serving as a kind of shorthand; but to those who believe that there is something real in it, this something is not very far from the *blas* of Van Helmont. *Vulcans*, hidden in the muscle or the nerve, are here detected by attraction, there by the production and the propagation of the nervous influx; that is to say, by phenomena of which we as yet know no analogues in the physical world, but of which we cannot say that they do not exist.

X. Bichat and G. Cuvier: Vital and Physical Properties Antagonistic.—The archeus and the blas of Van Helmont were but a first rough outline of vital properties. Xavier Bichat, the founder of general anatomy, wearied of all these incorporeal entities, of these unsubstantial principles with which biology was encumbered, undertook to get rid of them by the methods of the physicist and the chemist. The physics and the chemistry of his day referred phenomenal manifestations to the properties of matter, gravity, capillarity, magnetism, etc. Bichat did the same. He referred vital manifestations to the properties of living tissues, if not, indeed, of living matter. Of these properties as yet but very few were known: the irritability described by Glisson, which is the excitability of current physiology; and the irritability of Haller, which is nothing but muscular contractility. Others had to be discovered.

There is no need to recall the mistake made by Bichat and followed by most scientific men of his time, such as Cuvier in France, and J. Müller in Germany, for the story has been told by Claude Bernard. His mistake was in considering the vital properties not only as distinct from physical properties but even as opposed to them. The one preserve the body, the others tend to destroy it. They are always in conflict. Life is the victory of the one; death is the triumph of the other. Hence the celebrated definition given by Bichat: "Life is the sum total of functions which resist death," or the definition of the Encyclopædia: "Life is the contrary of death."

Cuvier has illustrated this conception by a graphic picture. He represents a young woman in all the health and strength of youth suddenly stricken by death. The sculptural forms collapse and show the angularities of the bones; the eyes so lately sparkling become dull; the flesh tint gives place to a livid pallor; the graceful suppleness of the body is now rigidity, "and it will not be long before more horrible changes ensue; the flesh becomes blue, green, black, one part flows away in putrid poison, and another part evaporates in infectious emanations. Finally, nothing is left but saline or earthy mineral principles, all the rest has vanished." Now, according to Cuvier, what has happened?

These alterations are the effect of external agents, air, humidity, and heat. They have acted on the corpse just as they used to act on the living being; but before death their assault had no effect, because it was repelled by the vital properties. Now that life has disappeared the assault is successful. We know now that external agents are not the cause of these disorders. They are caused by the microbes of putrefaction. It is against *them* that the organs were struggling, and not against physical forces.

The mistake made by Bichat and Cuvier was inexcusable, even in their day. They were wrong not to attach the importance they deserved to Lavoisier's researches. He had asserted, apropos of animal heat and respiration, the identity of the action of physical agents in the living body and in the external world. On the other hand, Bichat, by a flash of genius, decentralized life, dispersing the vital properties in the tissues, or, as we should now say, in the living matter. It was from the comparison between the constitution and the properties of living matter and those of inanimate matter that light was to come.

§ 3. Scientific Neo-Vitalism.

We can now understand the nature of modern neo-vitalism. It borrows from its predecessor its fundamental principle—namely, the specificity of the *vital fact*. But this specificity is no longer *essential*, it is only *formal*. The difference between it and the physical fact grows less and almost vanishes. It consists of a diversity of mechanisms or executive agents. For example, digestion transforms the alimentary starch in the intestines into sugar; the chemist does the same in his laboratory, only he employs acids, while the organism employs special agents, ferments, in this case a diastase. It is a particular form of chemistry, but still it is a chemistry. That is how Claude Bernard looked at it. The vital fact was not fundamentally distinguished from the physico-chemical fact, but only in form.

This expurgated and accommodated vitalism (Claude Bernard pushed his concessions so far as to call his doctrine "physico-chemical vitalism") was revived a few years ago by Chr. Bohr and Heidenhain.

Other biologists, instead of attributing the difference

between the phenomena of the two orders to the manner of their occurrence, seem to admit the complete identity of the mechanisms. It is no longer then in itself, individually, that the vital act is particularized, but in the manner in which it is linked to others. The vital order is a series of physico-chemical acts realizing an ideal plan.

Neo-vitalism has therefore assumed two forms, one the more scientific and the other the more philosophical.

Chr. Bohr and Heidenhain.—Its scientific form was given to it by Chr. Bohr, an able physiologist at Copenhagen, and by Heidenhain, a professor at Breslau, who was one of the lights of contemporary German physiology. The course of their researches led these two experimentalists, working independently, to submit to fresh investigation the ideas of Lavoisier and those of Bichat, on the relation of physico-chemical forces to the vital forces.

It was by no means a question of a general inquiry, deliberately instituted with the object of discovering the part played respectively by physical and physiological factors in the performance of the various functions. Such an investigation would have taken several generations to complete. No; the question had only come up incidentally. Chr. Bohr had studied with the utmost care the gaseous exchanges which take place between the air and the blood in the lungs. The gaseous mixture and the liquid blood are face to face; they are separated by thin membrane formed of living cells. Will this membrane behave as an inert membrane deprived of vitality, and therefore obeying the physical laws of the diffusion of gases? Well! no. It does not so behave. The most careful measurements of pressures and of solubilities leave no doubt in this respect. The living elements of the pulmonary membrane must therefore intervene in order to disturb the physical phenomenon. Things happen as if the exchanged gases were subjected not to a simple diffusion, a physical fact obeying certain rules, but to a real secretion, a physiological or vital phenomenon, obeying laws which are also fixed, but different from the former.

On the other hand, Heidenhain was led about the same time to analogous conclusions with respect to the liquid exchanges which take place within the tissues, between the liquids (lymphs) which bathe the blood-vessels externally and the blood which those vessels contain. The phenomenon is very important because it is the prologue of the actions of nutrition and assimilation. Here again, the two factors of exchange are brought into relation through a thin wall, the wall of the blood-vessel. The physical laws of diffusion, of osmosis, and of dialysis, enable us to foretell what would take place if the vitality of the elements of the wall did not intervene. Heidenhain thought he observed that things took place otherwise. The passage of the liquids is disturbed by the fact that the cellular elements are alive. It assumes the characteristics of a physiological act, and no longer those of a physical act. Let us add that the interpretation of these experiments is difficult, and it has given rise to controversies which still persist.

These two examples, around which others might be grouped, have led certain physiologists to diminish the importance of the physical factors in the functional activity of the living being to the advantage of the physiological factors. It would therefore seem that the vital force, to use a rather questionable form of language, withdraws in a certain measure the organized being from the realm of physical forces—and this conclusion is one form of contemporary neo-vitalism.

§ 4. Philosophical Neo-Vitalism.

Contemporary neo-vitalism has assumed another form, more philosophical than scientific, by which it is brought closer to vitalism, properly so called. We should like to mention the experiment of Reinke,[1] in Germany. Reinke is a botanist of distinction, who distinguishes the speculative from the positive domain of science, and cultivates both with success.

His ideas are analogous to those of A. Gautier, of Chevreul, and of Claude Bernard himself. He thinks, with these masters, that the mystery of life is not to be found in the nature of the forces that it brings into play, but in the direction that it gives them. All these thinkers are struck by the order and the direction impressed upon the phenomena which take place in the living being, by their interconnection, by their apparent adaptation to an end, by the kind of impression that they give of a plan which is being carried out. All these reflections lead Reinke to attach great weight to the idea of a "directing force."

The physico-chemical energies are no doubt the only ones which are manifested in the organized being, but they are directed as a blind man is by his

[1] Reinke, *Die Welt als That*; Berlin, 1899.

guide. It seems as if a *double* accompanies them like a shadow. This intelligent guide of blind, material force is what Reinke calls a *dominant*. Nothing could be more like the blas and the archeus of Van Helmont. Material energies would thus be paired off with their blas, their dominants, in the living organisms. In them there would therefore be two categories of force: "material forces," or rather, material energies obeying the laws of universal energetics; and in the second place, intelligent "spiritual forces," the dominants. When the sculptor is working his marble, in every blow which elicits a spark there is something more than the strong force of the hammer. There is thought, the volition of the artist, which is realizing a plan. In a machine there is more than machinery. Behind the wheels is the object which the author had in view when he adjusted them for a determined end. The energies spent in action are regulated by the adjustment—that is to say, by the dominants due to the intellect of the constructor.

Thus it is in the living machine. The dominants in this case are the guardians of the plan, the agents of the aim in view. Some regulate the functional activity of the living body, and some regulate its development and its construction. Such is the second form, the philosophical form, extreme and teleological, of contemporary neo-vitalism.

CHAPTER IV.

THE MONISTIC THEORY.

Physico-chemical Theory of Life. — Iatro-mechanism. — Descartes, Borelli.—Iatro-chemistry.—Sylvius le Boë.—The Physico-chemical Theory of Life.—Matter and Energy.—Heterogeneity is merely the result of the arrangement or combination of homogeneous bodies.—Reservation relative to the world of thought.—The Kinetic Theory.

THE unicist or monistic doctrine gives us a third way of conceiving the functional activity of the living being, by levelling and blending its three forms of activity—spiritual, vital, and material. It was expressed in the seventeenth and eighteenth centuries in "iatro-mechanism" and "iatro-chemistry," conceptions to which have more recently succeeded the physico-chemical doctrine of life, and finally "current materialism."

Materialism is not only a biological interpretation; it is a universal interpretation applicable to the whole of nature, because it is based on a determinate conception of matter. Here we find ourselves confronted by the eternal enigma discussed by philosophers relative to this fundamental problem of force and matter. We know what answers were given to the problem by the Ionic philosophers—Thales, Democritus, Heraclitus, and Anaxagoras, who discarded the agency of every spiritual power external to matter. The explanation of the world, the explanation of life, were reduced to the play of physical or mechanical forces. Epicurus, a little later, maintained that the knowledge of matter and its different forms accounts for all phenomena, and therefore for those of life.

Descartes, sharply separating the metaphysical world—that is to say, the soul defined by its attribute, thought—from the physical or material world characterized by extension, practically came to the same conclusions as the materialists of antiquity. To him, as to them, the living body was a mere machine.

Iatro-mechanism. Descartes. Borelli.—This, then, is the theory of the iatro-mechanicians, of which we may consider Descartes the founder, instead of the Greek philosophers. These ideas held their own for two centuries, and were productive of such fruitful results in the hands of Borelli, Pitcairn, Hales, Bernoulli, and Boerhaave, as to justify the jest of Bacon that "the philosophy of Epicurus had done less harm to science than that of Plato." The iatro-mechanic school tenaciously held its own until Bichat came upon the scene.

Iatro-chemistry. Sylvius le Boë.—It was from a reaction against their exaggerations that Stahl created animism, and the Montpellier school created vitalism. We gather some idea of the extravagant character of their explanations by reading Boerhaave. To this celebrated doctor the muscles were springs, the heart was a pump, the kidneys a sieve, and the secretions of the glandular juices were produced by pressure; the heat of the body was the result of the friction of the globules of blood against the walls of the blood-vessels; it was greater in the lungs because the vessels of the lungs were supposed to be narrower than those of other organs. The inadequacy of these explanations suggested the idea of completing them by the aid of the chemistry which was then springing into being. This chemistry, rudimentary as it was, longed for a share in the government of living bodies and in the explanation of their phenomena. Distillations, fermentations, and effervescences are now seen to play their rôle, a rôle which was premature and carried to excess. Iatro-chemistry from the general point of view is only an aspect of iatro-mechanics; but it is also an auxiliary. Sylvius le Boë and Willis were its most eminent representatives. This theory remained in the background until chemistry made its great advance—that is to say, in the days of Lavoisier. After that, its importance has gradually increased, particularly in the present day. Nowadays, the general tendency is to regard the organic functional activity, or even morphogeny—*i.e.*, whatever there is that is most peculiar to and characteristic of living beings—as a consequence of the chemical composition of their substance. This is a point of capital importance, and to it we must recur.

The Physico-chemical Theory of Life.—Contemporary biological schools have made many efforts to secure themselves from any slips on the philosophical side. They have avoided in most cases the psychological problem; they have deliberately refrained from penetrating into the world of the soul. Hence, *the physico-chemical theory* of life has been built up free from spiritualistic difficulties and objections. But this prudence did not exclude the tendency. And there is no doubt, as Armand Gautier said, that "real science can affirm nothing, but it also can deny nothing outside observable facts;" and again, that "only a science progressing backwards can venture to assert that matter alone exists, and that its laws alone govern the world." It is none the less true that by establishing the continuity between inert matter and living matter, we thereby render probable the continuity between the world of life and the world of thought.

Matter and Energy.—Besides, and without any wish to enter into this burning controversy, it is only too evident that there is no agreement as to the terms that are used, and in particular as to "matter" and "laws of matter." It is not necessary to repeat that the geometrical mould in which Descartes cast his philosophy has long since been broken. The celebrated philosopher, in defining matter by one attribute—extension, does not enable us to grasp its activity, an activity revealed by all natural facts; and in defining the soul by thought alone, prevents us from seeking in it the principle of this material activity. This purely passive matter, consisting of extension alone, this *bare matter* was to Leibniz a pure concept. A philosopher of our own time, M. Magy, has called it a sensorial illusion. The bodies of nature exhibit to us *matter clad* with energy, formed by the indissoluble union of extension with an inseparable dynamical principle. The Stoics declared that matter is mobile and not immobile, active and not inert. Leibniz also had this in his mind when he associated it indissolubly with an active principle, an "entelechy." Others have said that matter is "an assemblage of forces," or with P. Boscovitch, "a system of indivisible points without extension, centres of force, in fact." Space would be the geometrical locus of these points.

In this conception the materialistic school finds the

explanation of all phenomenality. Physical properties, vital phenomena, psychical facts, all have their foundation in this immanent activity. Material activity is a minimum of soul or thought which, by continuous gradation and progressive complexity, without solution of continuity, without an abrupt transition from the homogeneous to the heterogeneous, rises through the series of living beings to the dignity of the human soul. The observation of the transitions, an imperfect tracing of the geometrical method of limits, thus enables us to pass from material to vital, and from thence to psychical activity.

Apparent Heterogeneity is the Result of the Arrangement or the Combination of Homogeneous Bodies.—In this system, material energy, life, soul would only be more and more complex combinations of the consubstantial activity with material atoms. Life appears distinct from physical force, and thought from life, because the analysis has not yet advanced far enough. Thus, glass would appear to the ancient Chaldeans distinct from the sand and salt of which they made it. In the same way, again, water, to modern eyes, is distinct from its constituents, oxygen and hydrogen. The whole difficulty is that of explaining what this "arrangement" of the elements can introduce that is new in the aspect of the compound. We must know what novelty and apparent homogeneity the variety of the combinations, which are only special arrangements of the elementary parts, may produce in the phenomena. But we do not know, and it is this ignorance which leads us to consider them as heterogeneous, irreducible, and distinct in principle. The vital phenomenon, the complexus of physico-chemical facts, thus appears to us essentially different from those facts, and that is why we picture to ourselves "dominants" and "directing forces" more or less analogous to the sidereal guiding principle of Kepler, which, before the discovery of universal attraction, regulated the harmony of the movements of the planets.

A Reservation relative to the Psychical Order.—The scientific mind has shown in every age a real predilection towards the mechanical or materialistic theory. Contemporary scientists as a whole have accepted it in so far as it blends the vital and the physical orders. Objections and contradictions are only offered in the realm of psychology. A. Gautier, for example, has contested with infinite originality and vigour the claims of the materialists who would reduce the phenomenon of thought to a material phenomenon. The most general characteristic of material phenomenality is—as we shall later see—that it may be considered as a mutation of energy—*i.e.*, it obeys the laws of energetics. Now thought, says A. Gautier, is not a form of material energy. Thought, comparison, volition, are not acts of material phenomenality; they are states. They are realities; they have no mass; they have no physical existence. They respond to adjustments, arrangements, and concerted groupings of material manifestations of chemical molecules. They escape the laws of energetics.

Kinetic Theory.—We shall lay aside for a moment this serious problem relative to the limits of the world of conscious thought and of the world of life. It is on the other side, on the frontiers of living and inanimate nature, that the mechanical view triumphs. It has furnished a universal conception agreeing with phenomena of every kind—viz., the kinetic theory, which ascribes everything in nature to the movements of particles, molecules, or atoms.

The living and the physical orders are here reduced to one unique order, because all the phenomena of the sensible universe are themselves reduced to one and the same mechanics, and are represented by means of the atom and of motion. This conception of the world, which was that of the philosophers of the Ionic school in the remotest antiquity, which was modified later by Descartes and Leibniz, has passed into modern science under the name of the kinetic theory. The mechanics of atoms ponderable or imponderable, would contain the explanation of all phenomenality. If it were a question of physical properties or vital manifestations, the objective world in final analysis would offer us nothing but motion. Every phenomenon would be expressed by an atomistic integral, and that is the inner reason of the majestic unity which reigns in modern physics. The forces which are brought into play by Life are no longer to be distinguished in this ultimate analysis from other natural forces. All are blended in molecular mechanics.

The philosophical value of this theory is undeniable. It has exercised on physical science an influence which is justified by the discoveries which it has suggested. But to biology, on the other hand, it has lent no aid. It is precisely because it descends too deeply into things, and analyzes them to the uttermost, that it ceases to throw any light upon them. The distance between the hypothetical atom and the apparent and concrete fact is too great for the one to be able to throw light on the other The vital phenomenon vanishes with its individual aspect; its features can no longer be distinguished.

Besides, a whole school of contemporary physicists (Ostwald of Leipzig, Mach of Vienna) is beginning to cast some doubt on the utility of the kinetic hypothesis in the future of physics itself, and is inclined to propose to substitute for it the theory of energetics. We shall see, in every case, that this other conception, as universal as the kinetic theory, *the theory of Energy*, causes a vivid light to penetrate into the depths of the most difficult problems in physiology.

Such are, with their successive transformations, the three principal theories, the three great currents between which biology has been tossed to and fro. They are sufficiently indicative of the state of positive science in each age, but one is astonished that they are not more so; and this is due to the fact that these conceptions are too general. They soar too high above reality. More characteristic in this respect will be particular theories of the principal manifestations of living matter, of its perpetuity by generation, of the development by which it acquires its individual form, on heredity. It is here that it is of importance to grasp the progressive march of science—that is to say, the design and the plan of the building which is being erected, "blindly, so to speak," by the efforts of an army of workers, an army becoming more numerous day by day.

CHAPTER V.

THE EMANCIPATION OF SCIENTIFIC RESEARCH FROM THE YOKE OF PHILOSOPHICAL THEORIES.

The excessive use of Hypothetical Agents in Physiological Explanations—§ 1. *Vital Phenomena in Fully-constituted Organisms*—Provisory Exclusion of the Morphogenic idea—The Realm of the Morphogenic Idea as the Sanctuary of Vital Force—§ 2. *The Physiological Domain properly so called*—Harmony and Connection of Phenomena—Directive Forces—Claude Bernard's Work—Exclusion of Vital Force, of Final Cause, of the "Caprice" of Living Nature—Determinism—The Comparative Method—Generality of Vital Phenomena—Views of Pasteur.

THE theories whose history we have just sketched in broad outline long dominated science and exercised their influence on its progress.

This domination has ceased to exist. Physiology has emancipated itself from their sway, and this, perhaps, is the most important revolution in the whole history of biology. Animism, vitalism, materialism, have ceased to exercise their tyranny on scientific research. These conceptions have passed from the laboratory to the study; from being physiological, they have become philosophical.

This result is the work of the physiologists of sixty years ago. It is also the consequence of the general march of science and of the progress of the scientific spirit, which shows a more and more marked tendency to separate completely the domain of facts from the domain of hypotheses.

Excessive Use of Hypothetical Agents in Physiological Explanations.—It may be said that in the early part of the nineteenth century, in spite of the efforts of a few real experimenters from Harvey to Spallanzani, Hales, Laplace, Lavoisier, and Magendie, the science of the phenomena of life had not followed the progress of the other natural sciences. It remained in the fog of scholasticism. Hypotheses were mingled with facts, and imaginary agents carried out real acts, in inexpressible confusion. The soul *(animism)*, the vital force *(vitalism)*, and the final cause *(finalism, teleology)* served to explain everything.

In truth, it was also at this time that physical agents, electric and magnetic fluids, or, again, chemical affinity, played an analogous part in the science of inanimate nature. But there was at least this difference in favour of physicists and chemists, that when they had attributed some new property or aptitude to their hypothetical agents they respected what they attributed. The physiological physicians respected no law, they were subject to no restraint. Their vital force was capricious; its spontaneity made anticipation impossible; it acted arbitrarily in the healthy body; it acted more arbitrarily still in the diseased body. All the subtlety of medical genius was called into play to divine the fantastic behaviour of the spirit of disease. If we speak here of physiologists and doctors alone and do not quote biologists, it is because the latter had not yet made their appearance as authorities; their science had remained purely descriptive, and they had not yet begun to explain phenomena.

Such was the state of things during the first years of the nineteenth century. It lasted, thanks to the founders of contemporary physiology—Claude Bernard in France, and Brücke, Dubois-Reymond, Helmholtz, and Ludwig in Germany—until a separation took place between biological research and philosophical theories. This delimitation operated in physiology properly so called—*i.e.* in a branch of the biological domain in which as yet joint tenancy had been the rule. An important revolution fixed the respective divisions of experimental science and philosophical interpretation. It was understood that the one ends where the other begins, that the one follows the other, that one may not cross the other's path. There is between them only one doubtful region about which there is dispute, and this uncertain frontier is constantly being shifted and science daily gains what philosophy loses.

§ 1. Vital Phenomena in Constituted Organisms.

A displacement of this kind had taken place at the time of which we speak. It was agreed, that as far as concerns the phenomena which take place in *a constructed and constituted living organism*, it would no longer be permissible to allow to intervene in their explanation forces or energies other than those which are brought into play in inanimate nature. Just as when explaining the working of a clock, the physicist will not invoke the volition or the art of the maker, or the design that he had in view, but only the connection of causes and effects which he has utilized; so, for the living machine, whether the most complex, such as the human body, or the most elementary, such as the cell, we may not invoke a final cause, a vital force, external to that organism and acting on it from without, but only the connections and the fluctuations of effects which are the sole actual and efficient causes. In other words Ludwig, and Claude Bernard in particular, expelled from the domain of active phenomenality the three chimeras—Vital Force, Final Cause, and the "Caprice" of Living Nature.

But the living being is not only a *completely constructed and completely constituted* organism. It is not a finished clock. It is a clock which is making itself, a mechanism which is constructing and perpetuating itself. Nothing of the kind is known to us in inanimate nature. Physiology has found—in what is called morphogeny—its temporary limit. It is beyond this limit, it is in the study of phenomena by which the organism is constructed and perpetuated, it is in the region of the functions of generation and development, that philosophical doctrines expand and flourish. This is the present frontier of these two powers, philosophy and science. We shall presently delimit them more precisely. W. Kühne, a well-known scientist whose death is deplored, not in Germany alone, amused himself by studying the division of biological doctrines among the members of learned societies and in the world of academies. He summed up this kind of statistical inquiry by saying in 1898 at the Cambridge Congress, that physiologists were nearly all advocates of the physico-chemical doctrine of life, and that the majority of naturalists were advocates of vital force, and of the theory of final causes.

Domain of the Morphogenic Idea as the Last Sanctuary of Vital Force.—We see the reason for this. Physiology, in fact, has taken up its position in the explanation of the functional activity of the constituted organism—*i.e.*, on a ground where intervene, as we shall show further on, no energies and no matter other than universal energies and matter. Naturalists, on the other hand, have more especially considered—and from the descriptive point of view alone, at least up to the times of Lamarck and Darwin—the functions, the generation, the development and the evolution of species. Now these functions are most refractory and inaccessible to physico-chemical explanations. So, when the time came to give an account of what they had done, the zoologists had substituted for executive agents nothing but vital force under its different names. To Aristotle it is the vital force itself which, as soon as it is introduced into the body of the child, moulds its flesh and fashions it in the human form. Contemporary naturalists, the Americans C. O. Whitman and C. Philpotts, for example, take the same line of argument. Others, such as Blumenbach and Needham, in the eighteenth century, invoked the same division under another name, that of the *nisus formativus*. Finally, others play with words; they talk of heredity, of adaptation, of atavism, as if these were real, active, and efficient beings; while they are only appellations, names applied to collections of facts.

This region was therefore eminently favourable to the rapid increase of hypotheses, and so they abounded. There were the theories of Buffon, of Lamarck, of Darwin, of Herbert Spencer, of E. Haeckel, of His, of Weismann, of De Vries, and of W. Roux. Each biologist of any mark had his own, and the list is endless. But here already this domain of theoretical speculation is checked on various sides by experiment. J. Loeb, a pure physiologist, has recently given his researches a direction in which zoology believes may be found the explanation of the mysterious part played by the male element in fecundation. On the other hand, the first experiment of the artificial division of the living cell (*merotomy*), with its light upon the part played by the nucleus in the preservation and regeneration of the living form, is also the work of a physiological experimenter. It dates back to 1852, and is due to Augustus Waller. This experiment was made on the sensitive nervous cell of the spinal ganglions and on the motor cell of the anterior cornua of the spinal cord. The effects were correctly interpreted twelve or fifteen years later. All that zoologists have done is to repeat, perhaps unconsciously, this celebrated experiment and to confirm the result.

Thus we see that the attack upon the vitalist sanctuary has commenced. But it would be a grave mistake to suppose that final cause and vital force are on the point of being dislodged from their entrenchments. Philosophical speculation has an ample field before it. Its frontiers may recede. For a long time yet there will be room for a more or less modernized vitalism.

§ 2. THE PHYSIOLOGICAL DOMAIN PROPERLY SO CALLED.

Vitalism is even found installed in the region of physiology, although for the moment this science limits its ambition to the consideration of the completely constructed organized being, perfected in its form. The explanation of the working of this constituted machine cannot be complete until we take into account the harmony and the adjustment of its parts.

Harmony and Connection of Parts: Directive Forces.—These constituent parts are the cells. We know that the progress of anatomy has resulted in the cellular doctrine—*i.e.*, in the two-fold affirmation that the most complicated organism is composed of microscopic elements, the cells, all similar, true stones of the living building, and that it derives its origin from a single cell, egg, or spore, the sexual cell, or cell of germination. The phenomena of life, looked at from the point of view of the formed individual, are therefore harmonized in space; just as, regarded from the point of view of the individual in formation and in the species, they are connected in time. This harmony and this connection are in the eyes of the majority of men of science the most characteristic properties of the living being. This is the domain of *vital specificity*, of the *directive forces* of Claude Bernard and A. Gautier, and of the *dominants* of Reinke. It is not certain, however, that this order of facts is more specific than the other. Generation and development have been considered by many physiologists, and quite recently by Le Dantec, as simple aspects or modalities of nutrition or assimilation, the common and fundamental property of every living cell.

The Work of Claude Bernard. Exclusion of Vital Force, of Vital Cause, of the "Caprice" of Living Nature.—It is not, however, a slight advance or inconsiderable advantage to have eliminated vitalistic hypotheses from almost the whole domain of present-day physiology, and to have them, as it were, thrown back into its hinterland. This is the work of the scientific men of the first half of the nineteenth century, and particularly of Claude Bernard, who has thereby won the name of the founder or lawgiver of physiology. They found in the old medical school an obstinate adversary glorying in its sterile traditions. In vain was it proved that vital force cannot be an efficient cause; that it was a creation of the brain, an insubstantial phantom introduced into the anatomical marionette and moving it by strings at the will of any one—its adepts having only to confer upon it a new kind of activity to account for the new act. All that had been shown with the utmost clearness by Bonnet of Geneva, and by many others. It had also been said that the teleological explanation is equally futile, since it assigns to the present, which exists, an inaccessible, and evidently ultimately inadequate cause, which does not yet exist. These objections were in vain.

Determinism.—And so it was not by theoretical arguments that the celebrated physiologist dealt with his adversaries, but by a kind of lesson on things. In fact he was continually showing by examples that vitalism and the theory of final causes were idle errors which led astray experimental investigation; that they had prevented the progress of research and the discovery of the truth in every case and on every point in which they had been invoked. He laid down the principle of *biological determinism*, which is nothing but the negation of the "caprice" of living nature. This postulate, so evident that there was no need to enunciate it in the physical sciences, had to be shouted from the housetops for the benefit of supporters of vital spontaneity. It is the statement that, under determined circumstances materially identical, the same vital phenomena will be identically reproduced.

Comparative Method.—Claude Bernard completed this critical work by laying down the laws of experiment on living beings. He commended as the rational method of research the *comparative method.* This should be, and is in fact, the daily instrument of all those who work in physiology. It compels the investigator in every research bearing on organized beings to institute a series of tests, such that the conditions which are unknown and impossible to know may be regarded as identical from one test to another; and when we are certain that a single condition is variable, it compels him to discover the character of the condition we are dealing with, and to learn to appreciate, and to measure its influence. It is safe to say that the errors which are daily committed in biological work have their cause in some infraction of this golden rule. In physical science the obligation to follow the comparative method is much less felt. In most cases the *witness test*[1] is useless. In physiology the witness test is indispensable.

[1] In an article on the experimental method recently published in the *Dictionnaire de Physiologie*, M. Ch. Richet writes as follows:—"We must therefore never cease to carry out comparative experiments. I do not hesitate to say that this comparison is the basis of the experimental method." It is in fact what was taught by Claude Bernard in maxim and by example. It is no exaggeration to assert that nine-tenths of the errors which take place in research work are imputable to some breach of this method. When an investigator makes a mistake, save in the case of material error, it is almost certainly due to the fact that he has neglected to carry out one of the comparative

Generality of Vital Phenomena.—If we add that Claude Bernard opposed the narrow opinion, so dear to early medicine, which limited the consideration of vitality to man, and the contrary notion of the essential generality of the phenomena of life from man to the animal, and from the animal to the plant, we shall have given very briefly an idea of the kind of revolution which was accomplished about the year 1864, the date of the appearance of the celebrated *l'Introduction à la médecine expérimentale.*

The ideas we have just recalled seem to be as evident as they are simple. These principles appear so well founded that in a measure they form an integral part of contemporary mentality. What scientist would nowadays deliberately venture to explain some biological fact by the intervention of the evidently inadequate vital force or final cause? And who, to

tests required in the problem before him. The following is an instance which happened since the above pages were written:—Several years ago a chemist announced the existence in the blood serum of a ferment, lipase, capable of saponifying fats—that is to say, of extracting from them the fatty acid. From this he deduced many consequences relative to the mechanism of fermentations. But on the other hand, it has been since shown (April 1902) that this lipase of the serum does not exist. How did the error arise? The author in question had mixed normally obtained serum with oil, and he had noted the acidification of the mixture; he assured himself of the fact by adding carbonate of soda. He saw the alkalinity of the mixture, serum + oil + carbonate of soda, diminish, and he drew the conclusion that the acid came from the saponified oil. He did not make the comparative test, serum + carbonate of soda. If he had done so, he would have ascertained that it also succeeded, and that therefore as the acid did not come from the saponification of the oil, since there was none, its production could not prove the existence of a lipase.

account for the apparent inconsistency of the result, would bring forward the "caprice" of living nature? And who again would openly dispute the utility of the comparative method?

What the physiologists of to-day, according to Claude Bernard, would no longer do, their predecessors would do, and not the least important of them. Longet, for example, at a full meeting of the Académie, apropos of recurrent sensibility, and Colin (of Alfort), communicating his statistical results on the temperature of two hearts, accepted more or less explicitly the indetermination of vital facts. And why confine our remarks to our predecessors? The scientists of to-day are much the same. So here again we see the reappearance of the phantom of the final cause in so-called scientific explanations. One fact is accounted for by the necessity of the self-defence of the organism; another by the necessity to a warm-blooded animal of keeping its temperature constant. Le Dantec has recently reproached zoologists for giving as an explanation of fecundation the advantage that an animal enjoys in having a double line of ancestors. We might as well say, as L. Errera has pointed out, that the inundations of the Nile occur in order to bring fertility to Egypt.

We must not therefore depreciate the marvellous work which has emancipated modern physiology from the tutelage of early theories. The witnesses of this revolution appreciated its importance. One of them remarked as follows, on the appearance of *l'Introduction à la médecine expérimentale*, which contained, however, only a portion of the theory:—"Nothing more luminous, more complete, or more profound, has ever been written upon the true principles of an art so difficult as that of experiment. This book is scarcely known because it is on a level to which few people nowadays attain. The influence it will have on medical science, on its progress, and on its very language, will be enormous. I cannot now prove my assertion, but the reading of this book will leave so strong an impression that I cannot help thinking that a new spirit will at once inspire these splendid researches." This was said by Pasteur in 1866. That is what he thought of the work of his senior and his rival, at the moment when he himself was about to inspire those "splendid researches" with the movement of reform, the importance and the consequences of which have no equivalent in the history of science. By their discoveries and their teaching, by their examples and their principles, Claude Bernard and Pasteur have succeeded in emancipating a portion of the domain of vital facts from the direct intervention of hypothetical agents and first causes. They were compelled, however, to leave to philosophical speculation, to directing forces, to animism, to vitalism, an immense provisory field, the field which corresponds to the functions of generation and of development, to the life of the species and to its variations. Here we find them again in various disguises.

BOOK II.

THE DOCTRINE OF ENERGY AND THE LIVING WORLD.

Summary: General Ideas of Life.—Elementary Life.—Chapter I. Energy in General.—Chapter II. Energy in Biology.—Chapter III. Alimentary Energetics.

GENERAL IDEAS OF LIFE. ELEMENTARY LIFE.

Life is the Sum-total of the Phenomena Common to all Living Beings. Elementary Life.—Living beings differ more in form and configuration than in their manner of being. They are distinguished more by their anatomy than by their physiology. There are, in fact, phenomena common to all, from the highest to the lowest. This is because there is that similar or identical foundation, that *quid commune* which has enabled us to apply to them the common name of "living beings." Claude Bernard gave to this sum-total of manifestations common to all (nutrition, reproduction) the name of *elementary life.* To him *general physiology* was *the study of elementary life;* the two expressions were equivalent, and they were equivalent to a longer formula which the illustrious biologist has given as a title to one of his most celebrated works—*The Study of the Phenomena Common to all Living Beings, Animals, and Plants.* From this point of view each being is distinguished from another being as a given *individual* and as a particular *species*; but all are in some way alike and thus resemble one another: common life, elementary life, the essential phenomena of life; it is *life itself.*[1]

The manifestations of life may therefore be regarded from the point of view of what is most general among them. As we go down the scale of anatomical organization, as we pass from apparatus (circulatory, digestive, respiratory, nervous) to the *organs* which compose them, from the organs to the *tissues*, and finally from the tissues to the *anatomical elements* or *cells* of which they are formed, we approach that common, physiological dynamism which is *elementary life*, but we do not actually reach it. The cell, the anatomical element, is still a complicated structure. The elementary fact is further from us and lower down. It is in the living matter, in the molecule of this matter, and there we must seek it.

Galen gave in days gone by as the object of researches on life, the knowledge of the use of the different organs of the animal machine; "de usu partium." Later, Bichat assigned to them as their end the determination of the *properties of tissues.* Modern anatomists and zoologists try to reach the constituent element of these tissues—the cell. Their dream is to construct a *cellular physiology*, a *physiological cytology*; but we must go further than that.

[1] Le Dantec has objected to this conception of phenomena common to different living beings. He insists that all phenomena which take place in a given living being are proper to him, and differ, however slightly, from those of another individual. The objection is more specious than real.

General Physiology, Cellular Physiology, the Energetics of Living Beings.—General physiology, as was taught by Pflüger and his school, claims to go deeper down than the apparatus, or the organ, or even the cell. As in the case of physics, general physiology endeavours to reach, and really does in many cases reach, as far as the molecule. It is not cellular, it is *molecular.* Already, in fact, the efforts of modern science have succeeded in penetrating into the most general phenomena of the living being—those attributable to living matter, or, to speak more clearly, those which result from the play of the universal laws of matter at work in this particular medium which is the organized being.

Robert Mayer and Helmholtz have the honour of having set physiology in the right road. They founded *the energetics of living beings—i.e.*, they regarded the phenomena of life from the point of view of energy, which is the factor of all the phenomena of the universe.

CHAPTER I.

ENERGY IN GENERAL.

Origin of the Idea of Energy.—A new term, namely *energy*, has been for some years introduced into natural science, and has ever since assumed a more and more important place. It is owing to the English physicists, and especially to the English electrical engineers, that this expression has made its way into technology, an expression which is part and parcel of both languages, and which has the same meaning in both. The idea it expresses is, in fact, of infinite value in industrial applications, and that is why its use has gradually spread and become generalized. But it is not merely a practical idea. It is above all a theoretical idea of capital importance to pure theory. It has become the point of departure of a science, *energetics*, which, although born but yesterday, already claims to embrace, co-ordinate, and blend within itself all the other sciences of physical and living nature, which the imperfection of our knowledge alone had hitherto kept distinct and apart.

On the threshold of this new science we find inscribed *the principle of the conservation of energy*, which has been presented to us by some as Nature's supreme law, and which we may say dominates natural philosophy. Its discovery marked a new era and accomplished a profound revolution in our conception of the universe. It is due to a doctor, Robert Mayer, who practised in a little town in Wurtemberg, and who formulated the new principle in 1842, and afterwards developed its consequences in a series of publications between 1845 and 1851. They remained almost unknown until Helmholtz, in his celebrated memoir on the conservation of force, brought them to light and gave them the importance they deserved. From that time forward the name of the doctor of Heilbronn, until then obscure, has taken its place among the most honoured names in the history of science.[1]

[1] Mayer's claim to fame has been disputed. A Scotch physicist, P. G. Tait, has investigated the history of the law of the conservation of energy, which is the history of the idea of energy. The conception has taken time to penetrate the human mind, but its experimental proof is of recent date. P. G. Tait finds an almost complete expression of the law of the conservation of energy in Newton's third law of motion—namely, "the law of the equality of action and reaction," or rather, in the second explanation which Newton gave of that law. In fact, it was from this law that Helmholtz deduced it in 1847. He showed that the law of the equality of action and reaction, considered as a law of nature, involved the impossibility of perpetual motion, and the impossibility of perpetual motion is, in another form, the conservation of energy.

At a meeting of the Academy of Science, at Berlin, 28th March 1878, Du Bois-Reymond violently attacked Tait's contention. The honour of having been the first to conceive of the idea of energy and conservation was awarded to Leibniz. Newton had no right to it, for he appealed to divine intervention

As for *energetics*, of which thermodynamics is only a section, it is agreed that even if it cannot forthwith absorb mechanics, astronomy, physics, chemistry, and physiology, and build up that general science which will be in the future the one and only science of nature, it furnishes a preparation for that ideal state, and is a first step in the ascent to definite progress.

Here I propose to expound these new ideas, in so far as they contain anything universally accessible; and in the second place, I propose to show their application to physiology—that is to say, to point out their rôle and their influence in the phenomena of life.

Postulate: the Phenomena of Nature bring into play only two Elements, Matter and Energy.—If we try to account for the phenomena of the universe, we must admit with most physicists that they bring into play two elements, and two elements only namely, *matter* and *energy*. All manifestations are exhibited in one or other of these two forms. This, we may say, is the postulate of experimental science.

Just as gold, lead, oxygen, the metalloids, and the metals are different kinds of matter, so it has been recognized that sound, light, heat, and generally, the imponderable agents of the days of early physics, are

to set the planetary system on its path when disturbed by accumulated perturbations. On the other hand, Colding claims to have drawn his knowledge of the law of conservation from d'Alembert's principle. Whatever may be the theoretical foundations of this law, we are here dealing with its experimental proof. According to Tait, the proof can no more be attributed to R. Mayer than to Seguin. The real modern authors of the principle of the conservation of energy, who gave an experimental proof of it, are Colding, of Copenhagen, and Joule, of Manchester.

different varieties of energy. The first of these ideas is older and more familiar to us, but it has not for that reason a more certain existence. Energy is objective reality for the same reason that matter is. The latter certainly appears more tangible and more easily grasped by the senses. But, upon reflection, we are assured that the best proof of their existence, in both cases, is given by the law of their conservation—that is to say, their persistence in subsisting.

The objective existence of matter and that of energy will therefore be taken here as a postulate of physical science. Metaphysicians may discuss them. We have but little room for such a discussion.

§ 1. MATTER.

It is certainly difficult to give a definition of matter which will satisfy both physicists and metaphysicians.

Mechanical Explanation of the Universe. Matter is Mass.—Physicists have a tendency to consider all natural phenomena from the point of view of mechanics. They believe that there is a mechanical explanation of the universe. They are always on the look out for it, implicitly or explicitly. They endeavour to reduce each category of physical facts to the type of the facts of mechanics. They have made up their minds to see nowhere anything but the play of motion and force. Astronomy is celestial mechanics. Acoustics is the mechanics of the vibratory movements of the air or of sonorous bodies. Physical optics has become the mechanics of the undulations of the ether, after having been the mechanics of emission — a wonderful mechanics which represents exactly all the phenomena of light, and furnishes us with a perfect objective image of it. Heat, in its turn, has been reduced to a mode of motion, and thermodynamics claims to embrace all its manifestations. As early as 1812, Sir Humphry Davy wrote as follows:—"The immediate cause of heat is motion, and the laws of transmission are precisely the same as those of the transmission of motion." From that time forth, this conception developed into what is really a science. The constitution of gases has been conceived by means of two elements—particles, and the motions of these particles, determined in the strictest detail. And finally, in spite of the difficulties of the representation of electrical and magnetic phenomena after Ampère and before Maxwell and Hertz, physicists have been able to announce in the second half of the nineteenth century the unity of the physical forces realized in and by mechanics. From that time forth, all phenomena have been conceived as motion or modes of motion, only differing essentially one from the other in so far as motions may differ—that is to say, in the masses of the moving particles, their velocities, and their trajectories. The external world has appeared essentially homogeneous; it has fallen a prize to mechanics. Above all, there is heterogeneity in ourselves. It is in the brain, which responds to the nervous influx engendered by the longitudinal vibration of the air, by the specific sensation of sound, which responds to the transverse vibration of the ether by a luminous sensation, and in general to each form of motion by an irreducible specific sensation.

Forty years have passed since the mechanical explanation of the universe reached its definite and perfect form. It dominates physics under the name of the *theory of kinetic energy.* The minds of men in our own time are so strongly impregnated with this idea that most scientists of ordinary culture get no glimpse of the world of phenomena but by means of this conception. And yet it is only an hypothesis. But it is so simple, so intuitive, and appears to be so thoroughly verified by experiment, that we have ceased to recognize its arbitrary and unnecessarily contingent character. Many physicists from this standpoint consider the kinetic theory as an imperishable monument.

However, as in the case of H. Poincaré, the most eminent physicists and mathematicians are not the dupes of this system; and without failing to recognize the immense services which it has rendered to science, they are perfectly well aware that it is only a system, and that there may be other systems. Certain among them, such as Ostwald, Mach, and Duhem, believe that the monument is showing signs of decay, and at present the theory is opposed by another theory—namely, the theory of *energy.*

The theory of *energy* is usually considered and presented as a consequence of the kinetic theory; but it is perfectly independent of it, and it is, in fact, without relying on the kinetic theory, without assuming the unity of physical forces, which are combined in molecular mechanics, that we shall expound the general system.

This is not the point at issue for the moment. It is not a question of deciding the reality or the merit of this or that mechanical explanation; it is a question of something more general, because upon it depends the *idea of matter.* It is a question of knowing if there are any explanations other than mechanical. The illustrious English physicist, Lord Kelvin, does not seem willing to admit this. "I am never satisfied," he said, in his *Molecular Mechanics*, "until I have made a mechanical model of the object. If I can make this model, I understand; if I cannot, I do not understand."

This tendency of so vigorous a mind to be content only with mechanical explanations, has been that of the majority of scientific men up to the present day, and from it has arisen the scientific idea of matter.

What is matter, in fact, to the student of mechanics? It is mass. All mechanics is constructed of masses and forces. Laplace said: "The mass of a body is the sum of its material points." To Poisson, mass is the quantity of matter of which a body is composed. Matter is therefore confused with mass. Now, mass is the characteristic of the motion of a body under the action of a given force; it defines obedience or resistance to the causes of motion; it is the *mechanical parameter*; it is the co-efficient proper to every mobile body; it is the first *invariant* of which a conception has been established by science.

In fact, the word matter appears to be used in other senses by physicists, but this is only apparently so. They have but broadened the idea of the mechanicians. They have characterized matter by the whole series of phenomenal manifestations which are *proportional to mass*, such as weight, volume, chemical properties—so that we may say that the notion of matter does not intervene scientifically with a different signification from that of mass.

Two kinds of Matter. Ponderable and Imponder-

able.—In physics we distinguish between two kinds of matter—ponderable, obeying the law of universal attraction or weight, and imponderable matter or ether, which we assume to exist and to escape the action of that force. Ether has no weight, or extremely little weight. It is material in so far as it has mass. It is its mass which confers existence on it from the mechanical point of view—a logical existence, inferred from the necessity of explaining the propagation of heat, light, or electricity.

It may be observed that the use of mass really comes to bringing another element, force, to intervene, and we shall see that force is connected with energy; thus it comes to defining matter indirectly by energy. The two fundamental elements are not therefore irreducible; on the contrary, they should be one and the same thing.

Energy is the only Objective Reality.—This fusion into one will become more evident still when we examine the different kinds of energy, each of which exactly corresponds to one of the aspects of active matter. Shall we define matter by *extension*, by the portion of space it occupies, as certain philosophers do? The physicist will answer that space is only known to us by the expenditure of energy necessary to penetrate it (the activity of our different senses). And then what is weight? It is *energy of position* (universal attraction). And so with the other attributes. So that if matter were separated from the energetic phenomena by means of which it is revealed to us—weight or energy of position, impenetrability or energy of volume, chemical properties or chemical energies, mass or capacity for kinetic energy—the very idea of matter would vanish. And that comes to saying that fundamentally there is only one objective reality, *energy*.

Philosophical Point of View.—But from the philosophical point of view are there objective realities? That is a wider question which throws doubt upon matter itself, and which it is not our place to investigate here. A metaphysician may always discuss and deny the existence of the objective world. It may be maintained that man knows nothing beyond his sensations, and that he only objectivates them and projects them outside himself by a kind of hereditary illusion. We must avoid taking sides in all these difficulties. Physics for the moment ignores them—*i.e.*, postpones their consideration.

In a first approximation we agree to consider ponderable matter only. Chemistry acquaints us with its different forms. They are the different simple bodies, metalloids, metals, and the compound bodies, mineral or organic. Hence we may say that chemistry is *the history of the transformations of matter*. From the time of Lavoisier this science has followed the transformations of matter, balance in hand, and ascertains that they are accomplished without change of weight.

Law of the Conservation of Matter.—Imagine a system of bodies enclosed in a closed vessel, and the vessel placed in the scale of a balance. All the chemical reactions capable of completely modifying the state of this system have no effect upon the scale of the balance. The total weight is the same before, during, and after. It is precisely this equality of weight which is expressed in all the equations with which treatises on chemistry are filled.

From a higher point of view we recognize here, in

this *law of Lavoisier* or of the *conservation of weight*, the verification of one of the great laws of nature which we extend to every kind of matter, ponderable or not. It is the *law of the conservation of matter*, or again, of the indestructibility of matter—"Nothing is lost, nothing is created, all is transformation." This is exactly what Tait held, this impossibility of creating or destroying matter which at the same time is a proof of its objective existence. This indestructibility of ponderable matter is at the same time the fundamental basis of chemistry. Chemical analysis could not exist if the chemist were not sure that the contents of his vessel at the end of his operations ought to be quantitatively, that is to say by weight, the same as at the beginning, and during the whole course of the experiment.[1]

§ 2. ENERGY.

The Idea of Energy Derived from the Kinetic Theory.—The notion of energy is not less clear than the notion of matter, it is only more novel to our minds. We are led to it by the mechanical conception which now dominates the whole of physics, *the kinetic conception*, according to which in the sensible universe there are no phenomena but those of motion. Heat, sound, light, with all their manifestations so complex and so varied, may, according to this theory, be explained by motion. But then, if outside the brain and the mind which has consciousness and which perceives, Nature really offers us only motion, it follows that all phenomena are essentially homo-

[1] It must be added that the absolute rigour of this law has been called in question in recent researches. It would only have an approximate value.

geneous among one another, and that their apparent heterogeneity is only the result of the intervention of our sensorium. They differ only in so far as movements are capable of differing—that is to say, in velocity, mass, and trajectory. There is something fundamental which is common to them and this *quid commune* is *energy*. Thus the idea of energy may be derived from the kinetic conception, and this is the usual method of exposition.

This method has the great inconvenience of causing an idea which lays claim to reality to depend upon an hypothesis. And besides that, it gives a view of it which may be false. It makes of the different forms of energy something more than varieties which are equivalent to one another. It makes of them *one and the same thing*. It blends into one the modalities of energy and mechanical energy. For the experimental idea of equivalence, the kinetic theory substitutes the arbitrary idea of the equality, the blending, and the fundamental homogeneity of phenomena. This no doubt is how the founders of energetics, Helmholtz, Clausius, and Lord Kelvin understood things. But a more attentive study and a more scrupulous determination not to go beyond the teaching of experiment should compel us to reform this manner of looking at it. And it is Ostwald's merit that, after Hamilton, he insisted on this truth—that the various kinds of physical magnitudes furnished by the observation of phenomena are different and characteristic. In particular, we may distinguish among them those which belong to the order of *scalar* magnitudes and others which are of the order of *vector* magnitudes.

The Idea of Energy derived from the Connection of Phenomena.—The idea of energy is not absolutely connected with the kinetic theory, and it should not be exposed therefore to the vicissitudes experienced by that theory. It is of a higher order of truth. We can derive it from a less unsafe idea, namely that of the *connection of natural phenomena.* To conceive it we must get accustomed to this primordial truth, that there are no *phenomena isolated* in time and space. This statement contains the whole point of view of energetics.

The physics of early days had only an incomplete view of things, for it considered phenomena independently the one of the other.

Phenomena for purposes of analysis were classed in separate and distinct compartments: weight, heat, electricity, magnetism, light. Each phenomenon was studied without reference to that it succeeded or that which should follow. Nothing could be more artificial than such a method as this. In fact, there is a sequence in everything, everything is connected up, *everything precedes and succeeds in nature*—in nature there are only series. The isolated fact without antecedent or consequent is a myth. Each phenomenal manifestation is in solidarity with another. It is a metamorphosis of one state of things into another. It is transformation. It implies a state of things anterior to that which we are observing, a phenomenal form which has preceded the form of the present moment.

Now there exists a link between the anterior state and the succeeding state—that is to say, between the new form which is appearing and the preceding form which is disappearing. The science of energy shows that something has passed from the first condition to the second, but covering itself as it were with a new garment; in a word, that something active and permanent subsists in the passage from one condition to another, and that what has changed is only the aspect, the appearance.

This constant something which is perceived beneath the inconstancy and the variety of forms, and which circulates in a certain manner from the antecedent phenomenon to its successor, is energy.

But still this is only a very vague view, and it may seem arbitrary. It may be made more exact by examples borrowed from the different categories of natural phenomena. There are energetic modalities in relation with the different phenomenal modalities. The different orders of phenomena which may be presented—mechanical, chemical, thermal, electrical—give rise to corresponding forms of energy.

When to a mechanical phenomenon succeeds a mechanical, thermal, or electrical phenomenon, we say, embracing transformation in its totality, that there has been a transformation of mechanical energy into another form of energy, mechanical, thermal, or electrical, etc.

This idea becomes more precise if we examine successively each of these cases and the laws which regulate them.

§ 3. Mechanical Energy.

Mechanical energy is the simplest and the oldest known.

Mechanical Elements: Time, Space, Force, Work. Power.—Mechanical phenomena may be considered under two fundamental conditions—*time* and *space*, which are, in a measure, logical elements, to which may be joined a third element, itself experimental, having its foundations in our sensations—namely, *force*, *work*, or *power*.

The ideas of force, work, and power, are drawn from the experience man has of his muscular activity. Nevertheless the greatest mathematical minds from from Descartes to Leibniz have been obliged to define and explain them clearly.

Force.—The prototype of force is weight, universal attraction. Experiment shows us that every body falls as long as no obstacle opposes its fall. This is so universal a property of matter that it serves to define it. The *force*, weight, is therefore the name given to the cause of the fall of the bodies.

Force in general is the *cause of motion*. Hence force exists only in so far as there is motion. There would be no force without action. This is Newton's point of view. It did not prevail, and was not the point of view of his successors. The name of force has been given not only to the cause which produces or modifies motion, but to the cause which resists and prevents it. And then not only have *forces in action* been considered (dynamics), but *forces at rest* (statics). Now, to Newton there was no statics. Forces do not continue to exist when they produce no motion; they are not in equilibrium, they are destroyed.

The idea of force therefore is a metaphysical idea which contains the idea of *cause*. But it becomes experimental immediately it is looked upon as resisting motion, according to the point of view of Newton's opponents. Its foundations lie in the muscular activity of man.

Man can support a burden without bending or

moving. This burden is a weight—that is to say, a mass acted on by the force of weight. Man resists this force so as to prevent its effect. If it were not annihilated by man's *effort*, this effect would be the motion or the fall of the heavy body. The *effort* and the force are thus in equilibrium, and the effort is equal and opposite to the force. It gives to the man who exercises it the conscious idea of *force*. Thus we know of force through effort. Every clear idea that we can have of *force* springs from the observation of our muscular effort.

The notion of force is thus an anthropomorphic notion. When an effect is produced in nature outside human intervention, we say that it is by something analogous to what in man is effort, and we give to this something a name which is also analogous, namely *force*. To give a name to *effort* and to compare efforts in magnitude, we need not know all about them, nor need we know in what they essentially consist, of what series of physical, chemical, and physiological actions they are the consequence. And so it is with force. It is a resistance to motion or the cause of motion. This cause of motion may be an anterior motion (kinetic force). It may be an anterior physical energy (physical and chemical forces).

Forces are measured in the C.G.S. system by comparing them with the unit called the Dyne.[1] In practice they are compared with a much larger unit—the gramme, which is the weight, the force acting on a unit of mass of one centimetre of distilled water at a temperature of 4° C.

[1] The dyne is the force which applied to the unit of mass produces a unit of acceleration.

Work.—The muscular activity of man may be brought into play in yet another manner. When we employ workmen, as Carnot said in his *Essai sur l'équilibre et le mouvement*, it is not a question of "knowing the burdens that they can carry without moving from their position," but rather the burdens that they can carry from one point to another. For instance, a workman may have to lift the water from the bottom of a well to a given height, and the case is the same for the animals we employ. "This is what we understand by force when we say that the force of a horse is equal to the force of seven men. We do not mean that if seven men were pulling in one direction and the horse in another that there would be equilibrium, but that in a piece of work the horse alone would lift, for example, as much water from the bottom of a well to a given height as the seven men together would do in the same time."[1]

Here, then, we have to do with the second form of muscular activity, which is called in mechanics, "work"—at least, if in the preceding quotation we attach no particular importance to the words "in the same time," and retain the employment of muscular activity only "under constant conditions." Mechanical work is compared with the elevation of a certain weight to a certain height. It is measured by the product of the force (understood in the sense in which it was used just now—that is to say, as causing or resisting motion) and the displacement due to this motion. The unit is the Kilogrammetre—that is to say, the work necessary to lift a weight of one kilogramme to the height of one metre.

[1] These words spoil the statement, for time has nothing to do with it.

It will be remarked that the idea of time does not intervene in our estimation of work. The notion of work is independent of the ideas of velocity and time. "The greater or less time that we take to do a piece of work is of no more assistance in measuring its magnitude than the number of years that a man may have taken to grow rich or to ruin himself can help to estimate the present amount of his fortune."

Going back to Carnot's comparison, an employer who employed his workmen only on piece-work,—that is to say, who would only care about the amount of work done, and would be indifferent to the time that they took over it,—would be at the same point of view as the advocates of the mechanical theory. M. Bouasse, whom we follow here, has remarked that this idea of mechanical work may be traced back to Descartes. His predecessors, and Galileo in particular, had quite a different idea of the way in which mechanical activity should be measured; and so, among the mathematicians of the eighteenth century, Leibniz and, later, John Bernoulli were almost alone in looking at it from this point of view.

Energy.—Work thus understood is *mechanical energy*. It represents the lasting and objective effect of the mechanical activity independent of all the circumstances under which it was carried out. The same work may be done under very different conditions of time, velocity, force, and displacement. It is therefore the permanent element in the variety of mechanical aspects. Work, for example, in the collision of bodies when the motion of a body appears to be destroyed on impact with another, reappears as indestructible *vis viva*. This, then, is exactly what we call *energy*; and if we agree to give it this name, we may say that the conservation of energy is invariable throughout all mechanical transformations.

Distinction between Work and Force, and between Energy and Work.—The history of mechanics shows us what trouble has been taken and what efforts have been made to distinguish work (now mechanical energy) from force.

It is worth while insisting on this distinction. It could be easily shown that force has no objective existence. It has no duration, no permanence. It does not survive its effect, motion. There is no conservation of force. It passes instantly from infinity to zero. It is a *vectorial magnitude*—that is to say, it involves the idea of direction. Work, on the other hand, is the real element; it is a *scalar magnitude* involving the idea of opposite directions, indicated by the signs + and −. Work and force are heterogeneous magnitudes. Energy, and this is the only characteristic by which it is distinguished from work, is an *absolute magnitude* to which we may not even give opposite signs.

An example may perhaps throw these characteristics into relief—namely, the hydraulic press. We have on the platform exactly the work which has been done on the other side. The machine has only made it change its form. On the contrary, the force has been infinitely multiplied. We may, in fact, consider an infinite number of surfaces equal to that of a small piston, placed and orientated at will within the liquid; each, according to Pascal's principle, will support a pressure equal to that which is exercised. As soon as we cease to support it, this infinity falls at once to zero.

Now what real thing could pass instantly from infinity to zero?

That skilful and very able physiologist, M. Chauveau, has endeavoured to use the same term *energy of contraction* for the two phenomena of effort (force) and work. It seems, however, from the point of view of the expenditure imposed on the organism, that these two modes of activity, *static contraction* (effort), and *dynamical contraction* (work), may be, in fact, perfectly comparable. But although this manner of conceiving the phenomena may certainly be exact, and may be of great value, the idea of force must none the less remain distinct from that of work. The persistence of the author in violating established custom in this connection has prevented him from enabling mechanieians and even some physiologists to understand and accept very useful truths.

Power.—The idea of mechanical *power* differs from those of force and work. The idea of time must intervene. It is not sufficient, in fact, in order to characterize a mechanical operation, to point to the task accomplished. It may be necessary or useful to know how much time it required. This is true, especially when we are concerned with the circumstances as well as the results of the performance of the work; and this is the case when we wish to compare machines. We say that the machine which does the work in the shortest space of time is the most powerful. The unit of power is the Kilogrammetre-second—that is to say, the power of a machine which does a kilogrammetre in a second. In manufactures we generally employ a unit 75 times greater than this—a *horse-power.* This is the power of a machine which does 75 kilogrammetres a second,

In the electrical industry we measure by *kilowatts*, which are equivalent to 1.36 horse power, or by a *watt*, a unit a thousand times smaller.

Let us add that the power of a machine is not an absolute and permanent characteristic of the machine. It depends on the circumstances under which the work is carried out, and that is why, in particular, we cannot appreciate the power of the human machine in comparison with industrial machines. Experience has shown that the mechanical power of living beings depends upon the nature of the work they are doing. In this connection we may mention some very interesting experiments communicated to the Institute, in the year VI., by the celebrated physicist, Coulomb. A man of the average weight of 70 kilogrammes was made to climb the stairs of a house 20 metres high. He ascended at the rate of 14 metres a minute, and he performed this daily task for four effective hours. This work was equivalent to 235,000 kilogrammetres. But if, instead of climbing without a burden, the same man had had to carry a load, the result would have been quite different. Coulomb's workman took up six loads of wood a day to a height of 12 metres in 66 journeys, corresponding to a maximum work of 109,000 instead of 235,000 kilogrammetres. The mechanical power of the human machine thus varied in the two cases in the ratio of 235 to 109.

The Two Aspects of Mechanical Energy: Kinetic and Potential.—Energy, or mechanical work, may present itself in two forms—kinetic energy, corresponding to the mechanical phenomenon which has really taken place, and *potential energy*, or the energy of reserve.

A body which has been raised to a certain height will, if it be let fall, perform work which can be exactly measured in kilogrammetres by the product of its weight into the height it falls. Such work may be utilized in many ways. In this way, for instance, public clocks are worked. Now, as long as the clock-weight is raised and not let go, and as long as it is motionless, the physics of early days would say that there is nothing to discuss; the phenomenon is the fall; it is going to take place, but at the present moment there has been no fall.

In energetics we do not reason in this way. We say that the body possesses a *capacity for work* which will be manifested when the opportunity arises, a storage of energy, a virtual or *potential energy*, or again, an *energy of position*, which will be transformed into actual energy or real work as soon as the body falls.

Let us ask whence this energy arises. It proceeds from the previous operation which has raised the weight from the surface of the soil to the position it occupies. For example, if it is a question of the weights of a public clock, which, by its fall, will develop in 15 days the work that is necessary to turn the wheels, to strike the bell, and to turn the hands, this work ought to bring to our minds the exactly equal and opposite work done by the clockmaker, who has to carry the clock-weight and to lift it up from the ground to its point of departure. The work of the fall is the faithful counterpart of the work of elevation. The phenomenon has therefore in reality two phases. We find in the second exactly what was put into the first, the same quantity of energy—*i.e.*, the same work,

Between these phases comes the intermediary phase of which we say that it is a period of virtual *or potential energy*. This is a way of remembering in some measure the preceding phenomenon—*i.e.*, the work of lifting up, and of indicating the phenomena which will follow—*i.e.*, the work of the fall. And thus we connect by our thoughts the present situation with the antecedent and with the consequent position, and it is from this consideration of continuity alone that the conception of energy springs—that is to say, of something which is conserved and is found to be permanent in the succession of phenomena. This energy of which we lose no trace does not appear to us new when it is manifested. Our imagination eventually materializes the idea of it. We follow it as a real thing, having an objective existence, which is asleep during the latent potential period, and is revealed or manifested later.

Among other examples, that of the coiled spring which is unwound is particularly suitable for showing this fundamental character of the idea of mechanical energy, an idea which is the clearest of all. Machines are only transformers and not creators of mechanical energy. They only change one form into another.

In the same way, too, a stream of water or the torrent of a mountainous region may be utilized for setting in motion the wheels and the turbines of the factories situated in the valley. Its descent produces the mechanical work which would be a creation *ex nihilo* if we do not connect the phenomenon with its antecedents. We look on it as a simple restitution, if we think of the origin of this water which has been transported and lifted in some way to its level by the play of natural forces—evaporation under the action of the sun, the formation of clouds, transport by winds, etc. And we here again see that a complex energy has been transformed, in its first phenomenal condition, into *potential energy*, and that this potential energy is always expended in the second phase without loss or gain.

The Diffeıent Kinds of Mechanical Energy; of Motion, of Position.—There are as many forms of energy as there are distinct categories of phenomena or of varieties in these categories. Physicists distinguish between two kinds of mechanical energy—energy of motion and energy of position. The energy of position presents several variants—energy of distance, which corresponds to force: of this we have just spoken; energy of surface, which corresponds to particular phenomena of surface tension; and energy of volume which corresponds to the phenomena of pressure. Energy of motion, *kinetic energy*, is measured in two ways: as work (the product of force and displacement, $W = fs$) or as *vis viva* (half the product of the mass into the square of the velocity $U = \frac{mv^2}{2}$.[1]

[1] We therefore notice that the measures of force and work bring in mass, space, and time. The typical force, weight, is given by $w = mg$. On the other hand, we have by the laws of falling bodies $v = gt$; $s = \frac{1}{2}gt^2$; whence $g = \frac{2s}{t^2}$; $w = m\frac{2s}{t^2}$; or, if F be the force, M the mass, L the space described, and T the time, we have $F = MLT^{-2}$, which expresses what are called the *dimensions* of the force—that is to say, the magnitudes with their degree, which enter into its expression. We may thus easily obtain the dimensions of work:—

$$Work = f \times s = \frac{mv^2}{2} = ML^2T^{-2}.$$

§ 4. THERMAL ENERGY.

In the elements of physics it is nowadays taught that mechanical work may be transformed into heat, and reciprocally that heat may be transformed into mechanical work. Friction, impact, pressure, and expansion destroy or annihilate the mechanical energy communicated to a body or to the organs of a machine. With the disappearance of motion we note the appearance of heat. Examples abound. The tyre of a wheel is heated by the friction of the road. Portions of steel are warmed by the impact with stone, as in the old flint and steel. Two pieces of ice were melted by Davy, who rubbed them one against the other, the external temperature being below zero. The boiling of a mass of water caused by a drill was noticed by Rumford in 1790, during the manufacture of bronze cannon. Metal, beaten on an anvil, is heated. A leaden ball flattened against a resisting obstacle shows increase of temperature carried to the point of fusion. Finally, and symbolically, we have the origin of fire in the fable of Prometheus, by rubbing together the pieces of wood which the Hindoos called *pramantha*. Correlation is constant between the thermal and mechanical phenomena, a correlation that becomes evident as soon as observers have ceased to restrict themselves to the determination in isolation of the one fact or the other. There is never any real destruction of heat and motion in the true sense of the word; what disappears in one form appears again in another; just as if something indestructible were appearing in a series of successive disguises. This impression is translated into words when we speak of the metamorphosis of mechanical into thermal energy.

The Mechanical Equivalent of Heat.—The interpretation assumes a remarkable character of precision, which at once strikes the mind when physics applies to these transformations the almost absolute accuracy of its measurements. We then find that the rate of exchange is invariable. Transformations of heat into motion, and of motion into heat, take place according to a rigorous numerical law, which brings into exact correspondence the quantity of each. Mechanical effect is estimated, as we have seen, by work, that is in kilogrammetres. Heat is measured in calories, the calorie being the quantity of heat necessary to raise from 0°C to 1°C a kilogramme of water (Calorie) or one gramme of water (calorie). It is found that whatever may be the bodies and the phenomena which serve as intermediaries for carrying out this transformation, we must always expend 425 kilogrammetres to create a Calorie, or expend 0·00234 Calories to create a kilogrammetre. The number 425 is the mechanical equivalent of the Calorie, or, as is incorrectly stated, of the heat. It is this constant fact which constitutes *the principle of the equivalence of heat and of mechanical work.*

§ 5. CHEMICAL ENERGY.

We cannot yet actually measure chemical activity directly, but we know that chemical action may give rise to all other phenomenal modalities. It is their most ordinary source, and it is to it that industries appeal to obtain heat, electricity, and mechanical action. In the steam engine, for instance, the work that is received arises from the combustion of carbon by the oxygen of the air. This gives rise to the heat which vaporizes the water, produces the tension of the steam, and ultimately produces the displacement of the piston. The theory of the steam engine might be reduced to these two propositions: chemical activity gives rise to heat, and heat gives rise to motion; or to use the language to which the reader by now will be accustomed, chemical energy is transformed into thermal energy, and that into mechanical energy. It is a series of phases and of instantaneous changes, and the exchange is always affected according to a fixed rate.

The Measurement of Chemical Energy.—Our knowledge of chemical energy is less advanced than that of the energies of heat and sensible motion. We have not yet reached the stage of numerical verifications. We can only therefore affirm the equivalence of chemical and thermal energies without the aid of numerical constants, because we do not yet, in the present state of science, know how to measure chemical energy directly. Other known energies are always the product of two factors: the mechanical energy of position, or work, is measured by the product of the force f, and the displacement s; work $=fs$; the mechanical energy of motion, $U=\frac{1}{2}mv^2$, is measured by the product of the mass into half the square of the velocity. Thermal energy is measured by the product of the temperature and the specific heat; electric energy by the product of the quantity of electricity (in coulombs) and of the electromotive force (in volts). As for chemical energy, we guess that it may be valued directly according to Berthollet's system, adopted by the Norwegian chemists, Guldberg and Waage, by means of the product of the masses and of a force, or co-efficient of affinity, which depends on the nature of the substances which are brought together, on the temperature, and on the other physical circumstances of the reaction. On the other hand, the researches of M. Berthelot enable us in many cases to obtain an indirect valuation in terms of the equivalent heat.

Its Two Forms.—It is interesting to note that chemical energy may also be regarded from the two states of *potential* and *kinetic energy*. The coal-oxygen system, to burn in the furnace of the steam engine, must be primed by preliminary work (local ignition), just as the weight that is raised and left motionless at a certain height requires a small effort to be detached from its support. When this condition is fulfilled, energy is at once manifest. We must admit that it existed in the latent state, in the state of *chemical potential energy*. Under the impulse received, the carbon combines with the oxygen and forms carbonic acid, $C + 2O$ becomes CO_2; potential energy is changed into actual chemical energy, and immediately afterwards into thermal energy. We should have only a very incomplete and fragmentary view of the reality of things if we were to consider this phenomenon of combustion in isolation. We must consider it in connection with what has actually created the energy which it is about to dissipate. This antecedent fact is the action of the sun upon the green leaf. The carbon which burns in the furnace of the machine comes from the mine in which it was stored in the form of coal—that is to say, of a product which was vegetable in its primitive form, and which was formed at the expense of the carbonic acid of the air. The plant had separated, at the expense of the solar energy, the carbon from the oxygen to which it was united in the carbonic acid of the atmosphere. It had created the system $C + 2O$. So that the solar energy produces the chemical potential energy which was so long before it was utilized. Combustion expends this energy in making carbonic acid over again.

Materialization of Energy.—The fertility of the idea of energy is therefore, as we see from all these examples, due to the relations it establishes between the natural phenomena of which it exhibits the necessary relation, destroyed by the excessive analysis of early science. It shows us that in the world of phenomena there is nothing but transformations of energy. And we regard these transformations themselves as the circulation of a kind of indestructible agent which passes from one form of determination to another, as if it were simply putting on a fresh disguise. If our intellect requires images or symbols to embrace the facts and to grasp their relation, it may introduce them here. It will materialize energy, it will make of it a kind of imaginary being, and confer upon it an objective reality. And for the mind, as long as it does not become the dupe of the phantom which it itself has created, this is an eminently comprehensive artifice which enables us to grasp readily the relations between phenomena and their bond of affiliation.

The world appears to us then, as we said at the outset, constructed with singular symmetry. It offers to us nothing but transformations of matter and transformations of energy; these two kinds of metamorphoses being governed by two laws equally inevitable, the conservation of matter and the conservation of energy. The first of these laws expresses the fact that matter is indestructible, and passes from one phenomenal determination to another at a rate of equivalence measured by weight; the second, that energy is indestructible, and that it passes from one phenomenal determination to another at a rate of equivalence fixed for each category by the discoveries of the physicists.

§ 6. TRANSFORMATIONS OF ENERGY.

The idea of energy has become the point of departure of a science, *Energetics*, to the establishment of which a large number of contemporary physicists, among whom are Ostwald, Le Châtelier, etc., have devoted their efforts. It is the study of phenomena, regarded from the point of view of *energy*. I have said that it claims to co-ordinate and to embrace all other sciences.

The first object of energetics should be the consideration of the different forms of energy at present known, their definition and their measurement. This is what we have just done in broad outline.

In the second place, each form of energy must be regarded with reference to the rest, so as to determine if the transformation of this into that is directly realizable, and by what means, and, finally, according to what rate of equivalence. This new chapter is a laborious task which would compel us to traverse the whole field of physics.

Of this long examination we need only concern ourselves here with three or four results which will be more particularly important in the case of applications to living beings. They refer to mechanical energy, to the relations of thermal energy and chemical energy, to the complete rôle of thermal energy, and finally to the extreme adaptability of electrical energy.

1. *Transformation of Mechanical Energy.*—Mechanical energy may change into every other form of energy, and all others can change into it, with but one exception, that of chemical energy. Mechanical effort does not produce chemical combination. What we know of the part played by pressure in the reactions of dissociation seems at first to contradict this assertion. But this is only in appearance. Pressure intervenes in these operations only as *preliminary work* or *priming*, the purpose of which is to bring the bodies into contact in the exact state in which they must be for the chemical affinities to be able to enter into play.

2. *Transformation of Thermal Energy; Priming.*—Thermal (or luminous) energy does not change directly into chemical energy. In fact, heat and light favour and even determine a large number of chemical reactions; but if we go down to the foundation of things we are not long before we feel assured that heat and light only serve in some measure for *priming* for the phenomenon, for preparing the chemical action, for bringing the body into the physical state (liquid, steam) or to the degree of temperature (400° C. for instance, for the combination of oxygen and hydrogen) which are the preliminary indispensable conditions for the entry upon the scene of chemical affinities.

On the contrary, chemical energy may really be transformed into thermal energy. We have an instance of this in the reactions which take place without the aid of external energy; and again, in those very numerous cases which, such as the combustion of hydrogen and carbon, or the decomposition of explosives, the reactions continue when once primed. I may make a further observation apropos of thermal and photic energy. These are not two really and essentially distinct forms, as was thought in the early days of physics. When we consider things objectively, there is absolutely no light without heat; light and heat are one and the same agent. According as it is at this or that degree of its scale of magnitude, it makes a stronger impression on the skin (sensation of heat) or on the retina (sensation of light) of man and animals. The difference may be put down to the diversity of the work and not to that of the agent. The kinetic theory shows us that the agent is qualitatively identical. The words heat and light only express the chance of the meeting of the radiant agent with a skin and a retina. At the lowest degree of activity this agent exerts no action on the terminations of the thermal cutaneous nerves, nor on the optic nerve-terminations. As this degree is raised the former of these nerves are affected (cold, heat) and are so to the exclusion of the nerves of vision. Then they are both affected (sensation of heat and light), and finally, beyond that, sight alone is affected. The transformation of one energy into the other is therefore here reduced to the possibility of increasing or decreasing the intensity of the action of this common agent in the exact proportions suitable for passing from one of the conditions to the other; and this is easy when it is a question of going up the scale in the case of light, and, on the contrary, it is not realizable directly, that is to say without external assistance, when it is a question of going down the scale again, in the case of heat.

3. *Heat a Degraded Form of Energy.*—We have seen that thermal energy is not directly transformed into chemical energy. There is yet another restriction in the case of this thermal energy if we study the laws which govern the circulation and the transformations of thermal energy; and the most important comes from the impossibility of transporting it from a body at a lower temperature to a body at a higher temperature. On the whole, and because of these restrictions, thermal energy is an imperfect variety of universal energy, or, as the English physicists call it, a degraded form.

4. *Simple Transformations of Electrical Energy. Its Intermediary Rôle.*—On the other hand, electrical energy represents a perfected and infinitely advantageous form of this same universal energy, and this explains the vast development of its industrial applications within less than a century. It is not that it is better known than the others in its nature and in the secret of its action. On the contrary, there is more dispute than ever as to its nature. To some, electricity, which is transported and propagated with the speed of light, is a real flux of the ether as was taught by Father Secchi, who compared it to a current of water in a pipe. It would do its work, just as the water of the mill does its work by flowing over a wheel or through a turbine. Electricity, like water in this case, would not be an energy in itself, but a means of transporting energy.

To others, such as Clausius, Hertz, and Maxwell, it is not so; the electric current is not a transport of energy. It is a state of the ether of a peculiar, specific kind, periodically produced (electric oscillation), and propagated with a speed of the order of that of light.

However that may be, what constitutes the essential peculiarity of electrical energy, and what causes its value, is that it is an incomparable agent of transformation. Every known form of energy may be converted into it, and inversely, electrical energy may be changed with the utmost facility into all other energies. This extreme adaptability assigns to it the part of an intermediary between the other less tractable agents. Mechanical energy, for instance, lends itself with difficulty to the production of light, that is to say, to a metamorphosis into photic energy (a variety of thermal energy). A fall of water cannot be directly utilized for lighting purposes. The mechanical work of this fall, which cannot be exploited in its present form, serves to set in motion in industrial lighting the installations, the electric machines, and the dynamos which feed the incandescent lamps. Mechanical work is changed into electrical energy, and it, in its turn, into thermal or photic energy. Electricity has here played the part of a useful intermediary.

The last part of energetics must be consecrated to the study of the general principles of this science. These principles are two in number, the principle of the *conservation of energy*, or Mayer's principle, and the principle of the transformation of energy, or Carnot's principle. The doctrine of energy thus reduces to two fundamental laws the multitude of laws, often known as "general," to which natural science is subject.

§ 7. THE PRINCIPLE OF THE CONSERVATION OF ENERGY.

In all that precedes, the principle of conservation has intervened at every step. In fact, the very idea of energy is connected with the existence of this principle. We first discover the idea in the work of the philosophical mathematicians who established the foundations of mechanics:—Newton, Leibniz, d'Alembert, and Helmholtz; or of inductive physicists such as Lord Kelvin. Its experimental proof, sketched by Marc Seguin and R. Mayer, is due to Colding and Joule.

It is Independent of the Kinetic Theory.—Mayer's law states that energy is indestructible; that all phenomenality is nothing but a transformation of energy from one form to another, and that this transformation takes place either at equal values, or rather, at a certain rate of equivalence. This is what takes place when thermal energy is transformed into mechanical energy (equivalent 425). This rate of equivalence is fixed by the researches of physicists for each category of energy.

It will be noticed that this law and this theory of energy, which is always presented by authors of elementary books as a consequence of the kinetic theory, is quite independent of it. In the preceding lines we have not even mentioned its name. We have not assumed that all phenomena are movements or transformations of movements, whether sensible or vibratory; we have not affirmed that what was passing from one phenomenal determination to another was the *vis viva* of the motion, as is the case in the impact of elastic bodies. No doubt the kinetic theory affords us a very striking image of these truths which are independent of it; but it may be false: and the theory of energy which assumes the minimum of possible hypotheses would yet be true.

It contains a great many other Principles.—The principle of the conservation of energy contains a large number of the most general principles of science. It may be shown without much difficulty that, for example, it contains the principle of the inertia of matter, laid down by Galileo and Descartes; that of the equality of action and reaction, due to Newton; and even that of the conservation of matter, or rather of mass, due to Lavoisier. And finally, it contains the experimental law of equivalence connected with the name of the English physicist Joule, from which may be derived the Law of Hess and the principle of the initial and final states which we owe to Berthelot.

It involves the Law of Equivalence.—Here we may be content with noticing that the law of the conservation of energy involves the existence of relations of equivalence between the different varieties. A certain quantity of a given energy, measured, as we have seen, by the product of two factors, is equivalent to a certain fixed quantity of quite a different form of energy into which it may be converted. The laws which govern energetic transformations therefore contain, from both the qualitative and the quantitative points of view, all the connections of the phenomena of the universe. To study these laws in their detail is the task that physics must take upon itself.

The conversion one into the other of the different forms of energy by means of equivalents is only a possibility. It is subject, in fact, to all sorts of restrictions, of which the most important are due to the second principle.

§ 8. Carnot's Principle. Its Generality.

The second fundamental principle is that of the transformations of equilibrium, or of the conditions of reversibility, or again, Carnot's principle. This principle, which first assumed a concrete form in thermodynamics, has been very widely extended. It has reached a degree of generality such that contemporary theoretical physicists such as Lord Kelvin, Le Chatelier, etc., consider it the universal law of physical, mechanical, and chemical equilibrium.

Carnot's principle contains, as was shown by G. Robin, d'Alembert's principle of virtual velocities, and according to physicists of to-day, as we have just remarked, it contains the laws peculiar to physico-chemical equilibrium. The application of this principle gives us the differential equations from which are derived numerical relations between the different energies, or the different modalities of universal energy.

Its Character.—It is very remarkable that we cannot give a general enunciation of this principle which by its revealing power has changed the face of physics. This is because it is less a law, properly so called, than a method or manner of interpreting the relations of the different forms of energy, and particularly the relations of heat and mechanical energy.

Conversion of Work into Heat and Vice-versâ.—The conversion of work into heat is accomplished without difficulty. For example, the hammering of a piece of iron on an anvil may bring it to a red heat. A shell which passes through an armour plate is heated, and melts and volatilizes the metal all round the hole it has made. By utilizing mechanical action under the form of friction all energy can be converted into heat.

The inverse transformation of heat into work, on the contrary, cannot be complete. The best motor that we can think of, and *à fortiori* the best we can realize, can only transform a third or a fourth of the heat with which it is supplied.

This is an extremely important fact. It is of incalculable importance to natural philosophy, and may be ranked among the greatest discoveries.

Higher and Degraded Forms of Energy.—Of these we may give an account by distinguishing among the forms of universal energy *higher forms*, and *lower* or *degraded forms*. Here we have the principle of the *degradation of energy* on its trial, and it may be regarded as a particular aspect of the second principle of energetics, or Carnot's principle. Mechanical energy is a higher form. Thermal energy is a lower form, a degraded form, and one which has degrees in its degradation. Higher energy, in general, may be completely converted into lower energy; for example, work into heat: the slope is easy to descend, but it is difficult to retrace our steps; lower energy can be only partially transformed into higher energy, and the fraction thus utilizable depends upon certain conditions on which Carnot's principle has thrown considerable light.

Thus, although in theory the thermal energy of a body may have its equivalent in mechanical energy, the complete transformation is only realizable from the latter to the former, and not from the former to the latter. This is due to a condition of thermal energy which is called *temperature*. The same quantity of thermal energy, of heat, may be stored in the same thermal body at different temperatures. If this quantity of thermal energy is in a very hot body we can utilize a large portion of it; if it is in a relatively cold body we can only convert a small portion of it into mechanical work. Thus the value of energy,—*i.e.*, its capacity of being converted into a higher and more useful form,—depends on temperature.

The Capacity of Conversion depends on Temperature.—The conversion of heat into work assumes two bodies of different temperatures, the one warm and the other cold; a boiler and a condenser. Every thermal machine conveys a certain amount of heat from the boiler to the condenser, and what is not thus carried is changed into work. This residue is only a small fraction, a quarter, or at most a third of the heat employed, and that, too, in the theoretically perfect machine, in the ideal machine.

This output, this utilizable fraction depends on the fall of temperature from the higher to the lower level, just as the work of a turbine depends on the height of the waterfall which passes through it. But it also depends on the conditions of this fall, on the accessory losses from radiation and conduction. However, Carnot has shown that the output is the same, and a maximum for the same fall of temperature, whatever be the working agent (steam, hot air, etc.), and whatever be the machine—provided that this agent, this substance which works is not exposed to accessory losses, that it is never in contact with a body having temperature different to its own—or again, that it is connected only with bodies impermeable to heat.

This is Carnot's principle in one of its concrete forms.

A machine which realizes this condition, that the agent (steam, alcohol, ether) is in relation, at all phases of its function, with bodies which can neither take heat from it nor give heat to it, is a *reversible machine.* Such a machine is perfect. The fraction of heat that it transforms into motion is constant; it is a maximum; it is independent of the motor, of its organs, of the agent: it accurately expresses the transformability of the heat agent into a mechanical agent under the given conditions.

The Degradation and Restoration of Energy.—The fraction not utilized, that which is carried to the condenser at a lower temperature, is *degraded.* It can only be used by a new agent, in a new machine in which the boiler has exactly the same temperature as the condenser in the first machine, and the new condenser has a lower temperature, and so on. The proportion of utilizable energy thus goes on diminishing. Its utilization requires conditions more and more difficult to realize. The thermal energy loses its potential and its value, and is further and further degraded as its temperature approaches that of the surrounding medium.

The degraded energy, theoretically, has kept its equivalent value but, practically, it is incapable of conversion. However, it is shown in physics that it can be raised and re-established at its initial level. But for that purpose another energy must be utilized and degraded for its benefit.

The End of the Universe.—What we have just

seen with respect to heat and motion is to some degree true of all other forms of energy, as Lord Kelvin has shown. The principle of the degradation of energy is very general. Every manifestation of nature is an energetic transformation. In each of these transformations there is a degradation of energy—*i.e.*, a certain fraction is lowered and becomes less easily transformable. So that the energy of the universe is more and more degraded; the higher forms are lowered to the thermal form, the latter increasing at temperatures which become more and more uniform. The end of the universe, from this point of view, would then be unity of (thermal) energy in uniformity of temperature.

Importance of the Idea of Energy in Physiology.—I have said that the application of Carnot's principle furnished numerical relations between the different energetic transformations.

The science of living beings has not yet reached that point of development at which it is possible for us to obtain its numerical relations. However, the consideration of energy and the principle of conservation has altered the outlook of physiology on many questions which are of the highest importance.

The determination of the sources from which plants and animals draw their vital energies; the mediate transformation of chemical energy into animal heat in nutrition, or into motion in muscular contraction; the chemical evolution of foods; the study of soluble ferments—all these questions are of considerable importance when we wish to understand the mechanisms of life. They are therefore departments of physiological energetics in which great advances have already been made.

CHAPTER II.

ENERGY IN BIOLOGY.

§ 1. Energy in Living Beings.—§ 2. The First Law of Biological Energetics:—All Vital Phenomena are Energetic Transformations.—§ 3. Second Law:—The Origin of Vital Energy is in Chemical Energy. Functional Activity and Destruction.—§ 4. Third Law:—The Final Form of Energetic Transformation in the Animal is Thermal Energy. Heat is an Excretum.

THE theory of energy was thought of and utilized in physiology before it was introduced into physics, in which it has exercised such an extraordinary influence. Robert Mayer was a physicist and a doctor. Helmholtz was equally at home in physiology and in physics. From the outset both had seen in this new idea a powerful instrument of physiological research. The volume in which Robert Mayer expounded, in 1845, his remarkable views on organic movement in relation to nutrition, and Helmholtz' commentary leave us in no doubt in this respect. The essay on the mechanical equivalent of heat, of a more particularly physical character, is six years later than the earlier work.

The Relations between Energetics and Biology.—The theory of energy is therefore only returning to its cradle; and to that cradle it returns with all the sanction of physical proof, as the most general theory ever proposed in natural philosophy, and the theory least encumbered with hypotheses. It reduces all particular laws to two fundamental principles—that of the conservation of energy, which contains the principles of Galileo and Descartes, of Newton, of Lavoisier, Joule's law, Hess's law, and Berthelot's principle of the initial and final states; and also Carnot's principle, from which are deduced the laws of physico-chemical and chemical equilibrium. These two principles therefore sum up the whole of natural science. They express the necessary relation of all the phenomena of the universe, their uninterrupted gentic connection, and their continuity.

A priori there would be little likelihood that a doctrine, so universal and so thoroughly verified in the physical world, could be restricted, and thus be useless to the living world. Such a supposition would be contrary to the scientific method, which always tends to the generalization and the explanation of elementary laws. The human mind has always proceeded thus: it has applied to the unknown order of living phenomena the most general laws of contemporary physics.

This application has been found legitimate, and has been justified by experiment whenever it has been a question of the laws or of the really fundamental or elementary conditions of phenomena. It has, on the other hand, however, been unfortunate when it has stopped short of secondary characteristics. When we now concede the subjection of living beings to these general laws of energetics, we are following a traditional method. There is no doubt that this application is legitimate, and that experiment will justify it *a posteriori*.

I will therefore grant, as a provisional *postulate*, the consequences of which will have to be ultimately justified, that the living and inanimate world alike show us nothing but *transformations of matter* and *transformations of energy*. The word phenomenon will have no other signification, whatever be the circumstances under which the phenomenon occurs. The varied manifestations which translate the activity of living beings thus correspond to transformations of energy, to conversions of one form into another, in conformity with the rules of equivalence laid down by the physicists. This conception may be formulated in the following manner:—*The phenomena of life have the same claim to be energetic metamorphoses as the other phenomena of nature.*

This postulate is the foundation of biological energetics. It may be useful to give some explanation relative to the signification, the origin, and the scope of this statement.

Biological energetics is nothing but general physiology reduced to the principles that are common to all the physical sciences. Robert Mayer and Helmholtz gave the best description of this science, and laid down its limits by defining it as "the study of the phenomena of life regarded from the point of view of energy."

§ I. Energy at play in Living Beings. Common or Physical Energies. Vital Energies.

Our first object will be to define and to enumerate the energies at play in living beings; to determine their more or less easy transformations from one to another, to bring to light the general laws which govern those transformations, and finally to apply them to the detailed study of phenomena. This programme may be divided into four parts.

In the physical world the specific forms of energy are not numerous. When we have mentioned mechanical, chemical, radiant (thermal and photic) energies, electrical energy, with which is blended magnetic energy, we have exhausted the catalogue of natural agents.

But is this list for ever closed? Are vital energies comprised in this list? These are the first questions which we must ask ourselves.

The iatro-mechanical school, on *a priori* grounds give an affirmative answer. No doubt there are in the living organism many manifestations which are pure physical manifestations of known energies, mechanical, chemical, thermal, etc. But are all the manifestations of the living being of this order? Are they all, henceforth, reducible to the categories and varieties of energy which are investigated in physics? This is the claim of the mechanical school. But the claim is rash. Our fundamental postulate affirms, in principle, that universal energy is manifested in living beings; but, as a matter of fact, there is no reason for the assertion that it does not assume particular forms, according to the circumstances peculiar to the conditions under which they are produced.

These *special forms of energy* manifested in the conditions suitable to living beings would swell the list drawn up by the physicists. And it would not be the first instance of an extension of this kind. The history of science records many remarkable cases. Scarcely a century has passed since we first heard of electrical energy. This discovery in the world of energy, which took place, so to speak, before our very eyes, of an agent which plays so large a part in nature, clearly leaves the door open to other surprises.

We shall therefore concede that there may be other forms of energy at work in living beings than those we already know in the physical world. This reservation would enable us to discover at once the essential characteristics by which vital phenomena are henceforth reduced to universal physics, and the purely formal differences still distinguishing them.

If there are really special energies in living beings, our monistic postulate leads us to assert that these energies are homogeneous with the others, and that they do not differ from them more than they differ among themselves. It is probable that some day they will be discovered external to living bodies, if the material conditions (which it is always possible to imagine) are realized externally to them. And if we must admit that the peculiarity of the medium is such that these forms must remain indefinitely peculiar to living beings, we may assert with every confidence that these special energies do not obey special laws. They are subject to the two fundamental principles of Robert Mayer and Carnot. They are exchanged according to fixed laws with the other physical forms of energies at present known.

To sum up, then, we must establish three categories in the forms of energy which express the phenomena of vitality.

In the first place, most of these energies are those

which have already been studied and recognized in general physics. They are the same energies: chemical, thermal, mechanical, with their characteristics of mutability, their lists of equivalents, and their actual and potential states.

In the second place, it may happen, and it probably will happen, as it happened in the last century in the case of electricity, that some new form of energy will be discovered belonging to the universal order as to the living order. This will be a conquest of general physics as well as of biology.

And finally we may rigorously and provisionally admit a last category of *vital energies properly so called.*

It is difficult to give much precision to the idea of *vital energies properly so called.*

It will be easier to measure them by means of equivalents than to indicate their nature. Besides, this is the ordinary rule in the case of physical agents. We can measure them, although we know not what they are.

Characteristics of Vital Energies.—We see why we cannot exhibit with precision, *a priori*, the nature of vital energies. In the first place, they are expressed by what takes place in the tissues in activity, and this cannot at present be identified with the known types of physical, chemical, and mechanical phenomena. This is a first, intrinsic reason for not being able to distinguish them readily, since what takes place is not distinguished by the phenomenal appearances to which we are accustomed.

There is a second, intrinsic reason. These vital phenomena are intermediary, as we shall see, between manifestations of known energies. They lie between a chemical phenomenon which always precedes them, and a thermal phenomenon which always follows them. They are lost sight of, as it were, between manifestations which strike our attention. Generally speaking, intermediary energies often escape us even in physics. Only the extreme manifestations are clearly seen. In the presence of the organism we are, as it were, in electric lighting works which are run by a fall of water, and at first we only see the mechanical energy of the falling water, of the turbine and dynamo at work, and the photic energy of the lamps which give the light. Electrical energy, an intermediary, which has only a transient existence, does not impose itself on our attention.

And so *vital energies* for this twofold reason, intrinsic and extrinsic, are not readily apparent. To reveal them, the careful analysis of the physiologists is required. They are acts, in most cases silent and invisible, which we should scarcely recognize but by their effects, after they have terminated in familiar, phenomenal forms. This is, for example, what goes on in the muscle in process of shortening, in the nerve carrying the nervous influx, in the secreting gland. And this is what constitutes the different forms of energy which we call *vital properties.* M. Chauveau and M. Laulanié use the phrase *physiological work* to distinguish them. *Vital energy* would be preferable. It better expresses the analogy of this special form with the other forms of universal energy; it helps us better to understand that we must henceforth consider it as exchangeable by means of equivalents with the energies of the physical world just as they are exchangeable one with another.

§ 2. First Law of Biological Energetics.

It is easy to understand, after these remarks, the significance and the scope of this assertion which contains the first principle of biological energetics—namely, that the phenomena of life have the same claim to be called energetic metamorphoses as the other phenomena of nature.

Irreversibility of Vital Energies.—However, there is one characteristic of vital energies which deserves the closest attention. Their transformations have a direction which is in some measure inevitable. They descend a slope which they never re-ascend. They appear to be irreversible. Ostwald has rightly insisted on this fundamental characteristic, which no doubt is not that of all the phenomena of the living being without exception, but which is certainly that of the most essential phenomena. There are reversible phenomena in organisms; there are energetic transformations which may take place from one form of energy to another, or *vice versâ.* But the most characteristic phenomena of vitality do not act in this way. We shall presently see that most functional physiological acts begin with chemical and end with thermal action. The series of energetic transformations takes place in an inevitable direction, from chemical to thermal energy. The order of succession of ordinary energies is thus determined in the machine of the organism, and therefore by the conditions of the machine. The order of transformation of vital energies is still more rigorously regulated, and the phenomena of life evolve from childhood to ripened years, and thence to old age, without a possible return.

The laws of biological energetics are three in number. First of all, there is the fundamental principle which we have just developed, and which is, so to speak, laid down *a priori*; and there are two other principles, those established by experiment and summing-up, as it were, the multitude of known physiological effects. Of these two experimental laws, one refers to the *origin* and the other to the *termination of the energies developed in living beings.*

§ 3. SECOND LAW OF BIOLOGICAL ENERGETICS.

The Oigin of Vital Energy.—Vital energies have their origin in one of the *external or common energies* —not in any one we choose, as might be supposed, but in one only: chemical energy. The third principle will show us that they terminate in another energy or a few others, also completely fixed.

It follows that the phenomena of life must appear to us to be a circulation of energy which, starting from one fixed point in the physical world, returns to that world by a few points, also fixed, after a transient passage through the animal organism.

Or more precisely, it is a transposition from the realm of matter into the world of energy, of the idea of the *vital vortex* of Cuvier and the biologists. They defined life by its most constant property—nutrition. Nutrition was exactly this current of matter which the organism obtains from without by alimentation, and which it throws out again by excretion; and the even momentary interruption of which, if complete, would be the signal of death. The cycle of energy is the exact counterpart of this cycle of matter.

The second truth taught us by general physiology, a truth which physiology learned from experiment, is enunciated as follows:—*The maintenance of life consumes none of its energy. It borrows from the external world all the energy which it expends, and borrows it in the form of potential chemical energy.* This is a translation into the language of energetics of the results acquired in animal physiology during the last fifty years. No comment is needed to exhibit the importance of such a truth. It reveals the origin of animal activity. It reveals the source from which proceeds that energy which at some moment of its transformations in the animal organism will be a *vital energy.*

The *primum movens* of vital activity is, therefore, according to this law, the chemical energy stored up in the immediate principles of the organism.

Let us try to follow, for a moment, this energy through the organism and to specify the circumstances of its transformations.

Organic Functional Activity, and the Destruction of Reserve-stuff.—Let us suppose then, for this purpose, that our attention is directed to a given limited part of this organism, to a certain tissue. Let us seize it, so to speak, by observation at a given moment, and let us make an examination of the functional activity starting from this conventional moment. This functional activity, like all other vital phenomena, will be the result, as we have just explained, of a transformation of the potential chemical energy contained in the materials held in reserve in the tissue. This is our first perceptible fact. This energy, when disengaged, will furnish to the vital action the means by which it may be prolonged.

There is, then, a *functional destruction.* There is, at

the beginning of the functional process, and by a necessary effect of that very process, a liberation of chemical energy; and that can only take place by a decomposition of the immediate principles of the tissue, or, as we may say, by a destruction of organic material. Claude Bernard insisted on this consideration, that the vital function is accompanied by a destruction of organic material. "When a movement is produced, when a muscle is contracted, when volition and sensibility are manifested, when thought is exercised, when a gland secretes, then the substance of the muscles, of the nerves, of the brain, of the glandular tissue, is disorganized, is destroyed, and is consumed." Energetics enables us to grasp the deeply-seated reason of this coincidence between chemical destruction and the functional activity, the existence of which Claude Bernard intuitively suspected. A portion of organic material is decomposed, is chemically simplified, becomes less complex, and loses in this kind of descent the chemical energy which it contained in its potential state. It is this energy which becomes the very texture of the vital phenomenon.

It is clear that the reserve of energy thus expended must be replaced, because the organism remains in equilibrium. Alimentation provides for this.

How does it provide for it? This is a question which deserves detailed examination. We cannot incidentally treat it in full; we can only indicate its main features.

How the supply of Reserve Stuff is kept up.—We know that food does not directly replace the reserve of energy consumed by the functional activity. It is not its potential chemical energy which replaces, purely and simply, the energy brought into play, consumed, or, better still, transformed in the active organ, or tissue. Food as it is introduced, inert food, does not, in fact, take up its place *as it is*, without undergoing changes in that organ and that tissue, in order to restore the *status quo ante*.

Before building up the tissue it will have undergone various modifications in the digestive apparatus. It will have also undergone changes in the circulatory apparatus, in the liver, and in the very organ we are considering. It is after all these changes that assimilation takes place. It will find its place and will have then passed into the state of *reserve*.

The food digested, modified, and finally incorporated as an integral part in the tissue in which it will be expended, is therefore in a new state, differing more or less from its state when it was ingested. It is a part of the living tissue in the state of constitutive reserve. Its potential chemical energy is not the same as that of the food introduced. It may differ from it very remarkably in consequence of sudden alterations.

We do not know for certain at the expense of what category of foods this or that given organ builds up its reserve stuff. There is a belief, for instance, according to M. Chauveau, that the muscle does its work at the expense of the reserve of *glycogen* which it contains. The potential chemical energy of this substance would be a source of muscular mechanical energy. But we do not know exactly at the expense of what foods, albumenoids, fats, or carbohydrates the muscle builds up the reserve of glycogen expended during its contraction. It is probable that it builds it up at the expense of each of the three categories after the various more or less simple alterations undergone by the materials in the digestive tube, the blood, the liver, or other organs.

This building up of reserve stuff, the complement and counterpart of *functional destruction*, is not chemical synthesis. It is, on the contrary, generally, and on the whole, a simplification of the food that has been introduced. This is true, at least as far as the muscle is concerned. However, to this operation, Claude Bernard has given the name of *organizing synthesis*, but the phrase is not a happy one. But in no case was the eminent physiologist deceived as to the character of the operation. "The organizing synthesis," says he, "remains internal, silent, hidden in its phenomenal expression, gathering together noiselessly the materials which will be expended."

These considerations enable us to understand the existence of the two great categories into which the eminent physiologist divides the phenomena of animal life: the phenomena of the *destruction of reserve-stuff* corresponding to *functional facts*—that is to say expenditures of energy; and the *plastic phenomena* of the *building-up of reserves* of organic regeneration, corresponding to *functional repose*—*i.e.*, to the supply of food to the tissues.

Distinction between Active Protoplasm and Reserve-stuff.—If it is not exactly in these terms that Claude Bernard formulated this fruitful idea, it is at any rate in this way that it is to be interpreted. This can be done by giving it a little more precision. We apply more rigorously than that great physiologist the distinction drawn by himself between *really active and living protoplasm* and the *reserve-stuff* which it prepares. To the latter is restricted the destruction by the functional activity and the building up by repose.

The classification of Claude Bernard is strictly true for reserve-stuff. It is easy to criticize the wavering and, as it were, dimly groping expressions in which the celebrated physiologist has shrouded his ideas. The old adage will excuse him: *Obscuritate rerum verba obscurantur.* In the depths of his ignorance he had a flash of genius; perhaps he did not find the definitive and, as it were, clearly-cut formula defining what was in his mind. But, in this respect, he has left his successors an easy task.

The Law of Functional Assimilation.—The progress of physiological knowledge compels us therefore to distinguish in the constitution of anatomical elements two parts—the materials of *reserve-stuff* and the *really active* and *living protoplasm.* We have just seen how the reserve-stuff behaves, alternately destroyed by functional activity, and built up afterwards by the ingestion of food, followed by the operations of digestion, elaboration, and assimilation. It remains to ask how this really living and protoplasmic matter behaves. Does it follow the same law? Is it destroyed during the functional activity, and is it afterwards replaced? As to this we can express no opinion. M. le Dantec fills a gap in our knowledge, in this respect, by an hypothesis. He assumes that this essentially active matter grows during functional activity, and is destroyed during repose. This is what he calls the *law of functional assimilation.* The protoplasm would therefore behave in an exactly contrary manner to the reserve-stuff. It will be its counterpart. But this is only an hypothesis which, in the present state of our knowledge, cannot be verified by experiment. We are at liberty to assert either that the protoplasm increases by functional activity or that it is destroyed. Neither the arguments nor the objections pro or con have any decisive value. The facts alleged on either side are capable of too many interpretations.[1]

The only favourable argument (not demonstrative) is furnished by energetics. It is this. The *re-building of the protoplasm* is not like the *organization of reserve-stuff*, a slightly complicated or even simplified phenomenon, as happens in the case of the reserve of muscular glycogen. The glycogen, in fact, is built up at the expense of foods chemically more complex. It is, on the contrary, a clearly synthetic phenomenon, certainly of chemical complexity, since it ends in building up the active protoplasm which is, in some measure, of the highest scale of complexity. Its formation at the expense of the simplest alimentary materials requires, therefore, an appreciable quantity of energy.

The assimilation which organizes the active protoplasm therefore requires energy for its realization. Now, at the moment of functional activity, and by a necessary consequence thereof, the chemical destruc-

[1] The reason is to be found in the large number of indeterminates in the problem we have to solve. It will be sufficient to enumerate them: the two substances which exist in the anatomical element, protoplasm and reserve-stuff, to which are attributed contrary rôles; the two conditions attributable to the protoplasm, of manifested or latent activity; the facultv possessed by both of being prolonged for an indeterminate period, and of encroaching each on its protagonist when its existence is at stake. Here are more elements than are necessary to explain the positive or negative results of all the experiments in the world.

tion or simplification of the substance of reserve takes place. Here is something that meets the case, and we may note the coincidence. It does not mean that the disposable energy is really used to increase the protoplasm, nor that the protoplasm itself is thereby increased. It merely signifies that the wherewithal exists to provide for that increase if it takes place.

It is therefore *possible* that the active protoplasm follows the law of functional assimilation; but it is *certain* that the reserve-stuff follows the law laid down by Claude Bernard.

All these considerations definitely result in the confirmation of this second law of general physiology, according to which all vital energies are borrowed from the potential chemical energy of the reserve-stuff of alimentary origin.

§ 4. The Third Law of Biological Energetics.

The third law of biological energetics is also drawn from experiment. It relates no longer to the point of departure of the cycle of animal energy, but to its final position. *The energetic transformations of the animal end in thermal energy.*

This is the most novel part of the theory, and, if we may say so, that least understood by physiologists themselves. The energy resulting from the chemical potential of food, having passed through the organism (or simply through the organ which we are considering in action), and having given rise to phenomenal appearances more or less diversified, more or less dim or clear, obscure or obvious, which are the characteristic or still irreducible manifestations of vitality, finally returns to the physical world. This return takes place (with certain exceptions which will be presently indicated) under the ultimate form of thermal energy. This we are taught by experiment. The phenomena of functional activity are exothermal.

Real vital phenomena thus lie between the chemical energy which gives rise to them, and the thermal phenomena to which they in their turn give rise. The place of the vital fact in the cycle of universal energy is therefore completely determined. This conclusion is of the utmost importance to biology. It may be expressed in a concise formula which sums up in a few words all that natural philosophy can teach as to energetics applied to living beings. "Vital energy is a transformation of chemical energy into thermal energy."

Exceptions.—There are some exceptions to the rigour of this statement, but they are not many in number. We must first of all remark that it applies to *animal life* alone.

In the case of vegetables, looked at as a whole, the law must be modified. Their vital energy has another origin, and another final form. Instead of being the destroyers of chemical potential energy, they are its creators. They build up by means of the inert and simple materials afforded them by the atmosphere and the soil, the immediate principles by which their cells are filled. Their vital functional activity forms by synthesis of the reserves, carbo-hydrates (sugars and starches), fats, albuminoid nitrogenous materials—that is to say, the same three principal categories of foods as those used by animals.

And to return to the latter, it should be observed that thermal energy is not the only final form of vital energy, as this dogmatic statement would have it supposed. It is only the principle of the final forms. The cycle of energy occasionally terminates in mechanical energy (phenomena of motion) and in a less degree in other energies; such as, for example, the electrical energy produced by the functional activity of the nerves and muscles in all animals, or in the functional activity of special organs in rays, torpedo-fish, and the malapterurus electricus, or finally, in the photic energy of phosphorescent animals. But these are secondary facts.

Heat is an Excretum.—The third principle of biological energetics may be therefore thus enunciated:—*Vital energy in its final form becomes thermal energy.* This principle teaches us that if chemical energy is the primitive generating form of vital energies, thermal energy is the form of waste, of emunctory, the degraded form as the physicists would say. Heat is in the dynamical order an excretion of animal life, as urea, carbonic acid and water, are excreta in the substantial order. By a false interpretation of the principle of the mechanical equivalence of heat, or through ignorance of Carnot's principle, certain physiologists have fallen into error when they still speak of the transformation of heat into motion or into into electricity in the animal organism. Heat is transformed into nothing in the animal organism. It is dissipated. Its utility arises not from its energetic value, but from the part it plays as a primer in the chemical reactions, as has been explained with reference to the general characteristics of chemical energy.

The Effect of Energetics on our Knowledge of the Relations of the Universe.—The consequences of these principles of energetic physiology, which give us so much and which are so clear, are of the greatest importance from the practical as well as from the theoretical point of view.

In the first place, they show us the position and the rank of the phenomena of life in the universe as a whole. They throw fresh light on the noble harmony of the animal and vegetable kingdoms which Priestley, Ingenhousz, Senebier, and the chemical school of the beginning of the nineteenth century discovered, and which was expounded by Dumas with incomparable lucidity and brilliance. Energetics is expressed in a line. "The animal world expends the energy accumulated by the vegetable world." It extends these views beyond the living kingdoms. It shows how the vegetable world itself draws its activity from the energy radiated by the sun, and how animals restore it again, in dissipated heat, to the cosmic medium. It extends the harmony of the two kingdoms to the whole of nature. The new science makes of the whole universe one connected system.

From a more limited point of view, and so that we may not restrict ourselves to a consideration of the domain of animal physiology, the laws of energetics sum up and explain a multitude of facts and of experimental laws—for example, the law of the intermittence of physiological activity, the facts of fatigue, the rôle and the general principles of alimentation, and the conditions of muscular contraction.

CHAPTER III.

ALIMENTARY ENERGETICS.

Various Problems of Alimentation. § 1. *Food the source of Energy and Matter.* The two forms of Energy afforded by Food—Vital Energy, Thermal Energy. Food the source of Heat. The rôle of Heat.—§ 2. *Measure of the output of Energy*—by the Calometric Method—by the Chemical Method.—§ 3. The regular type of Food, Biothermogenic, and the irregular type, Thermogenic.—§ 4. Food considered as the source of Heat. The Law of Surfaces. The limits of Isodynamics.—§ 5. Plastic rôle of Food. Preponderance of Nitrogenous Foods.

AMONG the problems on which energetics has thrown a vivid light we have mentioned alimentation, muscular contraction, and, more general still, the intermittence of vital functional activity. We shall begin with the study of alimentation.

The Different Problems of Alimentation.—What is a food? In what does alimentation consist? The dictionary of the *Académie* will give us our first answer. It tells us that the word food is applied to "every kind of matter, whatever may be its nature, which habitually serves or may serve for nutrition." This is very well put, but here again we must know what nutrition is, and that is not a simple matter; in fact, it practically means whatever is usually placed on the table in a civilized and polished society. But it is just the profound reasons for this traditional practice that we are trying to discover.

The problem of alimentation may be looked at in a thousand ways. It is culinary, no doubt, and gastronomic; but it is also economical and social, agricultural, fiscal, hygienic, medical, and even moral. But first and foremost, it is physiological. It comprises and assumes the knowledge of the general composition of foods, of their transformations in the digestive apparatus, and their comparative utility in the maintenance and the sound functional activity of the organism. To this first group of subjects for our discussion are attached others relating to the effects of inanition, of insufficient alimentation, and of over-feeding. And in order to throw light on all these aspects of the problem of alimentation, we have to lay bare the most intimate and delicate reactions by which the organism is maintained and recruited, and, in the words of a celebrated physiologist, "to penetrate into the kitchen of vital phenomena." And here neither Apicius, nor Brillat-Savarin, nor Berchoux, nor the moralists, nor the economists are of any use to us as guides. We must appeal to the scientists, who, following the example of Lavoisier, Berzelius, Regnault, and Liebig, have applied to the study of living beings the resources of general science, and have thus founded *chemical biology.*

This branch of science developed considerably in the second half of the nineteenth century. It has now its methods, its technique, its chairs at the universities, its laboratories, and its literature. It has particularly applied itself to the study of the "material changes" or the *metabolism* of living beings, and with that object in view it has done two things In the first place, it has determined the composition of the constituent materials of the organism; then analyzing qualitatively and quantitatively all that penetrates into that organism in a given time—that is to say, all the alimentary or respiratory ingesta, and all that issues from the organism, *i.e.*, all the excreta, all the *egesta*, —it has drawn up *nutritive balance sheets*, corresponding to the various conditions of life, whether naturally or artificially created. And thus we can determine the alimentary régimes which give too much, and which give too little, and which finally restore equilibrium.

We do not propose to give a detailed account of this scientific movement. This may be done in monographs. All we wish to indicate here is the most general result of these laborious researches—that is to say, the laws and the doctrines which are derived from them, and the theories to which they have given birth. It is by this alone that they are brought into relation with general science, and may therefore interest the reader. The facts of detail are never lacking to the historian; it is more profitable to show the movement of ideas. The theories of alimentation bring into conflict very different conceptions of the vital functional activity. And here we find a confused medley of opinions on which it is not without interest to endeavour to throw some light.

§ 1. Food, a Source of Energy and Matter.

Definitions of Food.—Before the introduction into physiology of the notion of energy, no one had succeeded in giving an exact idea and a precise definition of food and alimentation. Every physiologist and medical man who attempted it had failed, and this for various reasons.

The general cause of this failure was that most definitions, popular or technical, interposed the condition that the food must be introduced into the digestive apparatus. "It is," said they, "a substance which when introduced into the digestive tube undergoes, etc., etc." But plants draw food from the soil, and they possess no digestive apparatus; many animals have no intestinal tube; and in the case of certain rotifèra, the females possess a digestive apparatus, while the males have none. Nevertheless all animals feed.

On the other hand, there are other substances than those which use the digestive tract for the purpose of entering the organism, and which are eminently useful or necessary to the maintenance of life. In particular we may mention oxygen.

The distinctive feature of food is its *utility*—when conveniently introduced or employed—to the living being. Claude Bernard's definition is this:—A substance taken in the external medium "necessary for the maintenance of the phenomena of the healthy organism and for the reparation of the losses it constantly suffers." "A substance which supplies an element necessary for the constitution of the organism, or which *diminishes its disintegration*" (stored-up food); this is the definition of C. Voit, the German physiologist. M. Duclaux says, in his turn, but in far too general terms, that it is a substance which contributes to assure the sound functional activity of any of the organs of the living being. None of these ways of describing food gives a complete idea.

Food, the Source of Energy and Matter.—The intervention of the notion of energy enables us more completely to understand the true nature of food. We must, in fact, have recourse to the energetic conception if we desire to take into account all that the organism requires from food. It not only requires *matter*, but also, and most important of all, energy.

Investigators so far concentrated their thoughts exclusively on the necessity of a supply of matter—that is to say, they only looked upon one side of the problem. The living body presents, at each of its points, an uninterrupted series of disintegrations and reconstitutions, the materials being supplied from without by alimentation, and rejected by excretion. Cuvier gave to this unceasing circulation of ambient matter throughout the vital world the name of *vital vortex*, and he rightly saw in it the characteristic of nutrition, and the distinctive feature of life.

This idea of the *cycle of matter* has been completed in our own time by that of the *cycle of energy*. All the phenomena of the universe, and therefore those of life, are conceived of as energetic transformations. We now look at them in their relationship instead of considering them individually as of old. Each has an antecedent and a consequent unity with which it is connected in magnitude by the law of equivalents taught us by contemporary physics. And thus we may conceive of their succession as the cycle of a kind of indestructible agent, which changes only apparently, or assumes another form as it passes from one to the other, but its magnitude remains unaltered. This is energy. Thus, in the living being there is not only a circulation of matter, but also a circulation of energy.

The most general result of research in physiological

chemistry from the time of Lavoisier down to our own day has been to teach us that *the antecedent of the vital phenomenon is always a chemical phenomenon.* The vital energies are derived from the potential chemical energy accumulated in the immediate constituent principles of the organism. In the same way *the consequent phenomenon of the vital phenomenon is in general a thermal phenomenon.* The final form of vital energy is thermal energy. These three assertions as to the nature, the origin, and the final form of vital phenomena constitute the three fundamental principles, the three laws, of biological energetics.

Food, a Source of Heat. It is not quâ source of heat that food is the source of vital energy.—The place of vital energy in the cycle of universal energy is completely determined. It lies between the chemical energy which is its generating form and the thermal energy which is its form of disappearance, of breakdown, the "degraded form," as the physicists say. Hence we have a result which can be immediately applied in the theory of food—namely, that heat is in the dynamical order an excretum of the animal life rejected by the living being, just as in the substantial order, urea, carbonic acid and water, are the materials used up and again rejected by it. We therefore must not think of the transformation in the animal organism of heat into vital energy, as certain physiologists always do. Nor must we think, with Béclard, of its transformation into muscular movement; or, as others have maintained, into animal electricity. This is not only an error of doctrine but an error of fact. It proceeds from a false interpretation of the principle of the mechanical equivalent of heat and a misunderstanding of Carnot's principle. Thermal energy does not repeat the course of the energetic flux in the animal organism. The heat is not transformed into anything. It is simply dissipated.

The Part played by Animal Heat as a Condition of Physiological Manifestations.—Does this mean that heat is useless to life in the very beings in which it is most abundantly produced—*i.e.*, in man and in the warm-blooded vertebrates? So far from this being so, it is necessary to life. But its utility has a peculiar character which must neither be misunderstood nor exaggerated. It is not transformed into chemical or vital reactions, but merely creates for them a favourable condition.

According to the first principle of energetics, for the vital fact to be derived from the thermal fact, the heat must be preliminarily transformed into chemical energy, since chemical energy is necessarily an antecedent and generating form of vital energy. Now this regressive transformation is impossible according to the current theories of general physics. The part played by heat in the act of chemical combination is that of a primer to the reaction. It consists in placing the reacting bodies, by changing their state or by modifying their temperature, in the condition in which they ought to be for the chemical forces to come into play. For example, in the combination of hydrogen and oxygen by setting light to an explosive mixture, heat only acts as a primer to the phenomenon, because the two gases which are passive at ordinary temperatures, require to be raised to 400° C. before chemical affinity comes into play. And so it is with the reactions which go on in the organism. They have a maximum temperature, and the part played by animal heat is to furnish them with it.

It follows that heat intervenes in animal life in two capacities—first and foremost *as excretum*, or end of the vital phenomenon, of *physiological work*; and on the other hand, as a *condition* or *primer* of the chemical reactions of the organism; and generally, as a favourable condition for the appearance of the physiological manifestations of living matter. Thus, it is not dissipated in sheer waste.

I was led to adopt these views some years ago from certain experiments on the rôle played in food by alcohol. I did not then know that they had already been expressed by one of the masters of contemporary physiology, M. A. Chauveau, and that they were related in his mind to a series of conceptions and of researches of great interest, in the development of which I have since then taken a share.

Two Forms of Energy supplied to Animals by Food. —To say that food is simultaneously a supply of energy and a supply of matter, is really to express in a single sentence the fundamental conception of biology, in virtue of which life brings into play no substratum or characteristic dynamism. According to this, the living being appears to us as the seat of an incessant circulation of matter and energy, starting from the external world and returning to it. All food is nothing but this matter and this energy. All its characteristics, our views as to its rôle, its evolution, all the rules of alimentation are simple consequences of this principle, interpreted by the light of energetics.

And first of all, let us ask what forms of energy are afforded by food? It is easy to see that there are two—food is essentially a source of chemical energy; and secondarily and accessorily, it is a source of heat.

Chemical energy is the only energy, according to the second law of energetics, which may be transformed into vital energy. It is true at any rate for animals; for in plants it is otherwise. There the vital cycle has neither the same point of departure nor the same final position. The circulation of energy does not take place in the same manner.

On the other hand, and this we are taught by the third law, energy brought into play in vital phenomena is finally liberated and restored to the physical world in the form of heat. We have just said that this release of heat is employed in raising the temperature of the living being. It is animal heat.

Thus there are two forms of energy supplied by food, chemical and thermal.

It must be added that these are not the only forms, but the principal, and by far the most important. It is not absolutely true that heat is the only outcome of the vital cycle. It is only so in the subject in repose, contented to live idly without doing external mechanical work, without lifting a tool or a weight, even that of its own body. And again, speaking in this way, we neglect all the movements and all the mechanical work which is done without exercise of the volition, by the beating of the heart and of the arteries, the movements of respiration, and the contractions of the digestive tube.

Mechanical work is, in fact, another possible termination of the cycle of energy. But there is no longer anything necessary or inevitable in this, since motion and the use of force are in a certain measure subordinated to the capricious volition of the animal.[1]

[1] There is another reason why the rôle of mechanical energy, compared with that of thermal energy, is reduced, in the partition of afferent, alimentary energy—at least, in animals which have not

At other times, again, it is an electrical phenomenon which terminates the vital cycle, and it is, in fact, in this way that things happen in the functional activity of the nerves and muscles in all animals, and in the functional activity of the electrical organ in fish, such as the ray and the torpedo. Finally, the termination may be a photic phenomenon, and this is what happens in phosphorescent animals.

It is idle to diminish the power of these principles by proceeding to enumerate the whole of the exceptions to their validity. We know perfectly well that there are no absolute principles in nature. Let us say, then, that the energy which temporarily animates the living being is furnished to it by the external world under the exclusive form of potential chemical energy; but that, if there is only one door of entry, there are two exits. It may return to the external world in the principal form of thermal energy and in the accessory form of mechanical energy.

§ 2. MEASUREMENT OF THE SUPPLY OF ALIMENTARY ENERGY.

Calorimetric Method.—From what has preceded it is clear that if the energetic flux which circulates through the animal emerges, *in toto*, in the state of heat, the measurement of this heat becomes the

to do excessive work. The unit of heat, the Calorie, is equivalent to 425 units of work—*i.e.*, to 425 kilogrammetres. In the animal at rest, the number of kilogrammetres representing the different quantities of work done is small, the number of corresponding Calories is 425 times smaller. It becomes almost negligeable in comparison with the considerable number of Calories dissipated in the form of heat.

measurement of the vital energy itself, for the origin of which we must go back to the food. If the flux is divided into two currents, mechanical and thermal, they must both be measured and the sum of their values taken. If the animal does not produce mechanical work, and all ends in heat, we have only to capture, by means of a calorimeter, this energetic flux as it emerges, and thus measure in magnitude and numerically the energy in motion in the living being. Physiologists use for this purpose various types of apparatus. Lavoisier and Laplace used an ice calorimeter—that is to say, a block of ice in which they shut up a small animal, such as a guinea-pig; they then measured its thermal production by the quantity of ice it caused to melt. In one of their experiments, for instance, they found that a guinea-pig had melted 341 grammes of ice in the space of ten hours, and had therefore set free 27 Calories.

But since those days more perfect instruments have been invented. M. d'Arsonval employed an air calorimeter, which is nothing but a differential thermometer very ingeniously arranged, and giving an automatic record. Messrs. Rosenthal, Richet, Hirn and Kaufmann, and Lefèvre have used more or less simplified or complicated air calorimeters. Others, following the example of Dulong and Despretz, have used calorimeters of air and mercury, or with Liebermester, Winternitz, and J. Lefèvre (of Havre), have had recourse to baths. Here, then, there is a considerable movement of research which has led to the discovery of very interesting facts.

Measurement of the Supply of Alimentary Energy by the Chemical Method.—We may again reach our result in another way. Instead of surprising the current of energy as it emerges and in the form of heat, we may try and capture it at its entry in the form of potential chemical energy.

The evaluation of potential chemical energy may be effected with the same unit of measurement as the preceding—that is to say, the Calorie. If we consider man and mammals, for example, we know that there is only apparently an infinite variety in their foods. We may say that they feed on only three substances. It is a very remarkable fact that all the complexity and multiplicity of foods, fruits, grains, leaves, animal tissues, and vegetable products of which use is made, reduce to so great a simplicity and uniformity, that all these substances are of three types only: albuminoids, such as albumen or white of egg—foods of animal origin or varieties of albumen; carbo-hydrates, which are more or less disguised varieties of sugar; and finally, fats.

Here, then, from the chemical point of view, leaving out certain mineral substances, are the principal categories of alimentary substances. Here, with the oxygen that is brought in by respiration, is everything that penetrates the organism.

And now, what comes out of the organism? Three things only, water, carbonic acid, and urea. But the former are the products of the combustion of the latter. If we consider an adult organism in perfect equilibrium, which varies throughout the experiment neither in weight nor in composition, we may say that the receipts balance the expenditure. Albumen, sugar, fat, plus the oxygen brought in, balance quantitatively the water, carbonic acid, and urea expelled. Things happen, in fact, as if the foods of the three categories were burned up more or less completely by the oxygen.

It is this combustion that we have known since the days of Lavoisier to be the source of animal heat. We can easily determine the quantity of heat left by albumen passing into the state of urea, and by the starch, the sugars, and the fats reduced to the state of water and carbonic acid. This quantity of heat does not depend on the variety of the unknown intermediary products which have been formed in the organism. Berthelot has shown that this quantity of heat which measures the chemical energy liberated by these substances is identical with the quantity obtained by burning the sugar and the fats in a chemical apparatus, in a calorimetric bomb, until we get carbonic acid and water, and by burning albumen till we get urea. This result is a consequence of Berthelot's *principle of initial and ifinal states.* The liberated heat only depends on the initial and final states, and not on the intermediary states. The heat left in the economy by the food being the same as that left in the calorimetric bomb, it is easy for the chemist to determine it. It has thus been discovered that one gramme of albumen produces 4.8 Calories, one gramme of sugar 4.2 Calories, and one gramme of fat 9.4 Calories. We thus gather what a given ration—a mixture in certain proportions of these different kinds of foods—supplies to the organism and what energy it gives it, measured in Calories.

The calculation may be carried out to a high degree of accuracy if, instead of confining ourselves to the broad features of the problem, we enter into rigorous detail. It is only, in fact, approximately that we have reduced all foods to albumen, sugar, and fat, and all excreta to water, carbonic acid, and urea.

The reality is a little more complicated. There

are varieties of albumen, carbo-hydrates, and fatty bodies, the heats of combustion of which in the organism oscillate in the neighbourhood of the numbers 4.8, 4.2, and 9.4. Each of these bodies has been individually examined, and numerical tables have been drawn up by Berthelot, Rubner, Stohmann, Van Noorden, etc. The tables exhibit the thermal value or energetic value of very different kinds of foods.

In our climate, the adult average man, doing no laborious work, daily consumes a maintenance ration composed, as a rule, of 100 grammes of albuminoids, 49 grammes of fats, and 403 grammes of carbohydrates. This ration has an energetic value of 2,600 Calories.

It is therefore, thanks to the victories won in the field of thermo-chemistry, and to the principles laid down since 1864 by M. Berthelot, that this second method of attack on nutritive dynamism has been rendered possible. Physiologists, by the aid of these methods, have drawn up *balance-sheets of energy* for living beings just as they had previously established *balance-sheets of matter*.

Now, it is precisely researches of this kind that we have indicated here as a consequence of biological energetics, which in reality have helped to build up that principle. These researches have shown us that, in conformity with the *principles of thermodynamics*, there was not, in fact, in the organism, any transformation of heat into mechanical work, as the physiologists for a short time supposed, on the authority of Berthelot. With the help of our theory this mistake is no longer possible. The doctrine of energetics shows us in fact the current of energy dividing itself, as it issues from the living being, into two divergent branches, the one thermal and the other mechanical, external the one to the other although both issuing from the same common trunk, and having between them no relation but this, that the sum of their discharges represents the total of the energy in motion. Let us now translate these very simple notions into the more or less barbarous jargon in use in physiology. We shall be convinced as we go on of the truth of the saying of Buffon, that "the language of science is more difficult to learn than the science itself." We shall say, then, that chemical energy, that the unit of weight of the food which may be placed in the organism, constitutes the *alimentary potential*, the *energetic value* of this substance, its *dynamogenic power*. It is measured in units of heat, in Calories, which the substance may leave in the organism. The evaluation is made according to the principles of thermo-chemistry, by means of the numerical tables of Berthelot, Rubner, and Stohmann. The same number also expresses the *thermogenic power*, virtual or theoretical, of the alimentary substance. This energy being destined to be transformed into *vital energies* (Chauveau's *physiological work, physiological energy*), the dynamogenic or thermogenic value of the food is at the same time its biogenetic value. Two weights of different foods which supply the organism with the same number of Calories,—*i.e.* for which these numerical values are the same,—will be called *isodynamic* or *isodynamogenic*, *isobiogenetic*, *iso-energetic* weights. They will be equivalent from the point of view of their alimentary value. And finally, if, as is usually the case, the cycle of energy ends in the production of heat, the food which has been utilized for this purpose has a real *thermogenic value*, identical with its theoretical thermogenic value. In this case it might be determined experimentally by direct calorimetry, measuring the heat produced by the animal supposed absolutely unchanged and identical before and after the consumption of the food.

§ 3. Different Types of Foods. The Regular, Biothermogenic Type and the Irregular, Thermogenic Type.

Food is a source of thermal energy for the organism because it is decomposed within it, and undergoes within it a chemical degradation. Physiological chemistry tells us that whatever be the manner in which it is broken up, it always results in the same body and always sets free the same quantity of heat. But if the point of departure and the point of arrival are the same, it is possible that the path pursued is not constantly identical. For example, one gramme of fat will always give the same quantity of heat, 9.4 Calories, and will always come to its final state of carbonic acid and water; but from the fat to the mixture of carbonic acid gas and water there are many different intermediaries. In a word we get the conception of varied cycles of alimentary evolutions.

From the point of view of the heat produced it has just been said that these cycles are equivalent. But are they equivalent from the vital point of view? This is an essential question.

Let us imagine the most ordinary alternative. Food passes from the natural to the final state after being incorporated with the elements of the tissues, and after having taken part in the vital operations. The chemical potential only passes into thermal energy after having passed through a certain intermediary phase of vital energy. This is the normal case, *the regular type of alimentary evolution.* It may be said in this case that the food has fulfilled the whole of its function, it has served for the vital functional activity before producing heat. It has been *biothermogenic.*

The irregular or pure thermogenic type.—And now let us conceive of the most simple *irregular or aberrant type.* Food passes from the initial to the final state without incorporation in the living cells of the organism, and without taking part in the vital functional activity. It remains confined in the blood and the circulating liquids, but it undergoes in the end, however, the same molecular disintegration as before, and sets free the same quantity of heat. Its chemical energy changes at once into thermal energy. Food is a *pure thermogen.* It has fulfilled only one part of its work. It has been of slight vital utility.

Does this ever occur in reality? Are there foods which would be only *pure thermogens*—that is to say, which would not in reality be incorporated with the living anatomical elements, which would form no part of them either in a state of provisory constituents of the living protoplasm, or in the state of reserve-stuff; which would remain in the internal medium, in the blood and the lymph, and would there undergo their chemical evolution? Or again, if the whole of the food does not escape assimilation, would it be possible for part to escape it? Would it be possible for one part of the same alimentary substance to be incorporated, and for the rest to be kept in the blood or the lymph, in the circulating liquids *ad limina corporis*, so to speak? In other words, can the same food be according to circumstances a *biothermogen* or a *pure thermogen?* Some physiologists—Fick of Wurzburg, for instance—have claimed that this is really the case for most nitrogenous elements, carbohydrates, and fats; all would be capable of evolving according to the two types. On the other hand, Zuntz and von Mering have absolutely denied the existence of the aberrant or pure thermogenic type. No substances would be directly decomposed in the organic liquids apart from the functional intervention of the histological elements. Finally, other authors teach that there is a small number of alimentary substances which thus undergoes direct combustion, and among them is alcohol.

Liebig's Superfluous Consumption.—Liebig's *theory of superfluous consumption* and Voit's *theory of the circulating albumen* assert that the proteid foods undergo partial direct combustion in the blood vessels. The organism only incorporates what is necessary for physiological requirements. As for the surplus of the food that is offered it, it accepts it, and, so to speak, squanders it; it burns it directly; and we have a "sumptuary" consumption, consumption *de luxe.*

In this connection arose a celebrated discussion which still divides physiologists. If we disengage the essential body of the discussion from all that envelops it, we see that it is fundamentally a question of deciding whether a food always follows the same evolution whatever the circumstances may be, and particularly when it is introduced in great excess.

Liebig thought that the superabundant part, escaping the ordinary process, was destroyed by direct combustion. He affirmed, for instance, that nitrogenous substances in excess were directly burned in the blood instead of passing through their usual cycle of vital operations. We might express the same idea by saying that they then undergo an accelerated evolution. Instead of passing through the blood in the anatomical element, to return in the dismembered form from the anatomical element to the blood, their breaking up takes place in the blood itself. They save a displacement, and therefore in reality remain external to the construction of the living edifice. Their energy, crossing the intermediary vital stage, passes with a leap from the chemical to the thermal form. Liebig's doctrine reduced to this fundamental idea deserved to survive, but mistakes in minor details involved its ruin.

Voit's Circulating Albumen.—A few years later C. Voit, a celebrated physiological chemist of Munich, revived it in a more extravagant form. He held that almost the whole of the albuminoid element is burned directly in the blood. He interpreted certain experiments on the utilization of nitrogenous foods by imagining that these substances when introduced into the blood were divided as a result of digestion into two parts: the one very small, which was incorporated with the living elements, and passed into the stage of *organized albumen*, the other, corresponding to the greater part of the alimentary albumen, remained mingled with the blood and lymph, and was subjected in this medium to direct combustion. This was *circulating albumen*. In this theory the tissues are almost stable; the organic liquids alone are subjected to oxydizing transformations, to nutritive metabolism. The accelerated evolution, which Liebig considered as an exceptional case, was to C. Voit the rule.

Current Ideas as to the Rôle of Foods.—The ideas of to-day are not those of Voit; but they do not, however, differ from them essentially. We no longer admit that the greater part of the ingested and digested albumen remains confined in the circulating medium external to the anatomical elements. It is held, with Pflüger and the school of Bonn, that it penetrates the anatomical element and is incorporated in it; but in agreement with Voit it is believed that a very small part is assimilated to the really living matter, to the protoplasm properly so called; the greater part is deposited in the cellular element as reserve-stuff. The material, properly so called, of the living machine does not undergo destruction and reparation as extensively as our predecessors supposed. There is no need for great reparation. On the contrary, the physiological activity consumes to a great extent the reserve-stuff. And the greater part of the food, after having undergone suitable elaboration, serves to replace the reserve-stuff destroyed in each anatomical element by the vital functional activity.

Experimental Facts. — Among the facts which brought physiologists of the school of Voit to believe that most foods do not get beyond the internal medium, there is one which may well be mentioned here. It has been observed that the consumption of oxygen in respiration increases notably (about a fifth of its value) immediately after a meal. What does this mean? The interval is too short for the digested alimentary substances to have been elaborated and incorporated in the living cells. It is supposed that an appreciable time is required for this complete assimilation. The products of alimentary digestion are therefore in all probability still in the blood, and in the interstitial liquids in communication with it. The increase of oxygen consumed would show that a considerable portion of these nutritive substances absorbed and passed into the blood would be oxydized and then and there destroyed. But this interpretation, however probable it may be, does not really fit in with the facts in such a way that we may consider it as proved. Certain experiments by Zuntz and Mering are opposed to the idea that combustion in the blood is easy. These physiologists injected certain oxydizable substances into the vessels without being able to detect any instantaneous oxidation. It is only fair to add that against these fruitless attempts other more fortunate experiments may be quoted.

Category of Purely Thermogenic Foods, with Accelerated Evolution. Alcohol. Acids of Fruits.— The accelerated evolution of foods—an evolution which takes place in the blood, that is to say outside the really living elements—remains, therefore, very uncertain as far as ordinary food is concerned. It has been thought that it was a little less uncertain as far as the special category of alcohol, acids of fruits, and glycerine is concerned.

Some authors consider these bodies as pure thermogens. When alcohol is ingested in moderate doses, they say that about a tenth of the quantity absorbed becomes fixed in the living tissues; the rest is "circulating alcohol." It is oxidized directly in the blood and in the lymph, without intervening in the vital functions other than by the heat it produces. From the point of view of the energetic theory these are not real foods, because their potential energy is not transformed into any kind of vital energy, but passes at once to the thermal form. On the other hand, other physiologists look upon alcohol as really a food. According to them everything is called a food which is transformed in the organism with the production of heat; and they measure the nutritive value of a substance by the number of Calories it can give up to the organism. So that alcohol would be a better food than carbohydrated and nitrogenous substances. A definite quantity of alcohol, a gramme for instance, is equivalent from the thermal point of view to 1.66 grammes of sugar, 1.44 of albumen, or 0.73 of fat. These quantities would be *isodynamic.*

Experiment has not entirely decided for or against this theory. However, the first tests have not been very favourable to it. The researches of C. von Noorden and his pupils, Stammreich and Miura, have clearly and directly established that alcohol cannot be substituted in a maintenance ration for an exactly isodynamic quantity of carbohydrates. If the substitution is effected, a ration only just capable of maintaining the organism in equilibrium becomes insufficient. The animal decreases in weight. It loses more nitrogenous matter than it can recover from its diet, and this situation cannot be sustained for long. On the other hand, the celebrated researches of the American physiologist, Atwater, would plead, on the contrary, in favour of almost isodynamic substitution. Finally, Duclaux has shown that alcohol is a real food, biothermogenic for certain vegetable organisms. But urea is also a food for *micrococcus ureæ*. It does not follow that it is a food for mammals. We have not reached the solution yet—*adhuc sub judice.*

Conclusion: The Energetic Character of Food.—To sum up we have confined ourselves, in what has been said, to the consideration of a single character of food, and really the most essential, its energetic character. Food must furnish energy to the organism, and for that purpose it is decomposed and broken up within it, and issues from it simplified. It is thus, for instance, that the fats, which from the chemical point of view are complicated molecular edifices, escape in the form of carbonic acid and water. And so it is with carbo-hydrates, starchy and sugary substances. This is because these compounds descend to a lower degree of complexity during their passage through the organism, and by this drop, as it were, they get rid of the chemical energy which they contained in the potential state. Thermo-chemistry enables us to deduce from the comparison of the initial and final states the value of the energy absorbed by the living being. This energetic, dynamogenic or thermogenic value, thus gives a measure of the alimentary capacity of the substance. A gramme of fat, for instance, gives to the organism a quantity of energy equivalent to 9.4 Calories; the thermogenic value of the albumenoids is 4.8 Calories. The thermogenic or thermal value of carbohydrates is less than 4.7 calories. This being so, we understand why the animal is nourished by foods which are products very high in the scale of chemical complexity.

§ 4. FOOD CONSIDERED EXCLUSIVELY AS SOURCE OF HEAT.

We have seen that food is, in the first place, a source of *chemical energy;* and, in the second place, a source of *vital energy*—finally, and consequently, a source of thermal energy. It is this last point of view which has exclusively struck the attention of certain physiologists, and hence has arisen a peculiar manner of conceiving the rôle of food. It consists in looking on food as a source of thermal energy.

This conception is easily applied to warm-blooded animals, but to them exclusively—and this is where it first fails. The animal is warmer than the environment in general. It is constantly giving out heat to it. To repair this loss of heat it takes in food in exact proportion to the loss it sustains. When it is a question of cold-blooded vertebrates, which live in water and in most cases have an internal temperature which is not distinguishable from that of the environment, we see less clearly the thermal rôle of food. It seems then that the production of heat is an episodic phenomenon, not existing for itself.

However that may be, food is in the second place a source of thermal energy for the organism. Can it be said, inversely, that every substance which we introduce into the economy, and which is there broken up and gives off heat, is a food? This is a moot point. We dealt just now with purely thermogenic foods. However, most physiologists are inclined to give a positive answer. In their eyes the idea of food cannot be considered apart from the fact of the production of heat. They take the effect for the cause. To these physiologists everything ingested is called food, if it gives off heat within the body.

To be heated by food is, indeed, an imperious necessity for the higher animals. If this need be not satisfied the functional activities become enervated; the animal falls into a state of torpor; and if it is capable of attenuated, of more or less latent, life it sleeps in a state of hibernation; but if it is not capable of this, it dies. The warm-blooded animal with a fixed temperature is so organized that this constancy of temperature is necessary to the exercise and to the conservation of life. To maintain this indispensable temperature there must be a continual supply of thermal energy. According to this, the necessity of alimentation is confused with the necessity of a supply of heat to cover the deficit which is due to the inevitable cooling of the organism. This is the point of view taken up by theorists, and we cannot say that they have no right to do so. We can only protest against the exaggeration of this principle, and the subordination of the other rôles of food to this single rôle as a thermogen. It is the magnitude of the thermal losses which, according to these physiologists, determines the need for food, and regulates the total value of the maintenance ration. From the quantitative view it is approximately true. From the qualitative point of view it is false.

Such is the theory opposed to the theory of chemical and vital energy. It has on its side a large number of experts, among whom are Rubner, Stohmann, and von Noorden. It has been defended in an article in the *Dictionnaire de Physiologie* by Ch. Richet and Lapicque. They hold that thermogenesis absolutely dominates the play of nutritive exchanges;

and it is the need for the production of heat that regulates the total demand for Calories which every organism requires from its ration. It is not because it produces too much heat that the organism gets rid of it peripherally: it is rather because it inevitably disperses it that it is adapted to produce it.

Rubner's Experiments.—This conception of the rôle of alimentation is based on two arguments. The first is furnished by Rubner's last experiment (1893). A dog in a calorimeter is kept alive for a rather long period (two to twelve days); the quantity of heat produced in this lapse of time is measured, and it is compared with the heat afforded by the food. In all cases the agreement is remarkable. But is it possible that there should be no such agreement? Clearly no, because there is a well-known regulating mechanism which always exactly proportions the losses and the gains of heat to the necessity of maintaining the fixed internal temperature. This first argument is, therefore, not conclusive.

The second argument is drawn from what has been called the *law of surfaces*, clearly perceived by Regnault and Reiset in their celebrated memoir in 1849, formulated by Rubner in 1884, and beautifully demonstrated by Ch. Richet. In comparing the maintenance rations for subjects of very different weights, placed under very different conditions, it is found that the food always introduces the same number of Calories for the same extent of skin—*i.e.*, for the same cooling surface. The numerical data collected by E. Voit show that, under identical conditions, warm-blooded animals daily expend the same quantity of heat per unit of surface—namely, 1.036 Calories per square yard. The average ration introduces exactly the amount of food which gives off sensibly this number of Calories. Now, this is an interesting fact, but, like the preceding, it has no demonstrative force.

Objections. The Limits of Isodynamism.—On the contrary, there are serious objections. The thermal value of the nutritive principles only represents one feature of their physiological rôle. In fact, animals and man are capable of extracting the same profit and the same results from rations in which one of the foods is replaced by an *isodynamic* proportion of the other two—that is to say, a proportion developing the same quantity of heat. But this substitution has very narrow limits. Isodynamism—that is to say, the faculty that food has of supplying *pro ratâ* its thermal values—is limited all round by exceptions. In the first place, there are a few nitrogenous foods that no other nutritive principle can supply; and besides, beyond this minimum, when the supply takes place, it is not perfect. Lying between the albuminoids and the carbohydrates relatively to the fats, it is not between these two categories relatively to nitrogenous substances. If the thermal power of food were the only thing that had to be considered in it, the isodynamic supply would not fail in a whole category of principles such as alcohol, glycerine, and the fatty acids. Finally, if the thermal power of a food is the sole measure of its physiological utility, we are compelled to ask why a dose of food may not be replaced by a dose of heat. External warming might take the place of the internal warming given by food. We might be ambitious enough to substitute for rations of sugar and fat an isodynamic quantity of heat-giving coal, and so nourish the man by suitably warming his room. In reality, food has many other offices to fulfil than that of warming the body and of giving it energy—that is to say, of providing for the functioual activity of the living machine. It must also serve to provide for wear and tear. The organism needs a suitable quantity of certain fixed principles, organic and mineral. These substances are evidently intended to replace those which have been involved in the cycle of matter, and to reconstitute the organic material. To these materials we may give the name of *histo-genetic* foods (repairing the tissues), or of *plastic* foods.

§ 5. THE PLASTIC RÔLE OF FOOD.

Opinions of the Early Physiologists.—It is from this point of view that the ancients regarded the rôle of alimentation. Hippocrates, Aristotle, and Galen believed in the existence of a unique nutritive substance, existing in all the infinitely different bodies that man and the animals utilize for their nourishment. It was Lavoisier who first had the idea of a dynamogenic or thermal rôle of foods. Finally, the general view of these two species of attributes and their marked distinction is due to J. Liebig, who called them *plastic* and *dynamogenic* foods. In addition he thought that the same substance should accumulate the same attributes, and that this was the case with the albuminoid foods, which were at once *plastic* and *dynamogenic*.

Preponderance of Nitrogenous Foods.—Magendie, in 1836, was the pioneer who introduced in this interminable list of foods the first simple division. He divided them into proteid substances, still called albuminoids, nitrogenous, quaternary, and *ternary substances.* Proteid substances are capable of maintaining life. Hence the preponderant importance given by the eminent physiologist to this order of foods. These results have since been verified. Pflüger, of Bonn, gave a very convincing proof of this a few years ago. He fed a dog, made it work, and finally fattened it, by giving it nothing at all to eat but meat from which had been extracted, as thoroughly as possible, every other substance.[1] The same experiment showed that the organism can manufacture fats and carbo-hydrates at the expense of the nitrogenous food, when it does not find them ready formed in the ration. The albumen will suffice for all the needs of energy and and matter. To sum up, there is no necessary fat, no carbohydrate is necessary; albuminoids alone are indispensable. Theoretically, the animal and man alike could maintain life by the exclusive use of proteid food; but, practically, this is not possible for man, because of the enormous amount of meat which would have to be used (3 kilogrammes a day).

Ordinary alimentation comprises a mixture of three orders of substances, and to this mixture albumen brings the plastic element materially necessary for the reparation of the organism ; it also is the source of energy. The two other varieties only bring energy. In this mixed regimen the quantity of albumen must never descend below a certain minimum. The efforts of physiologists of late years have tended to fix with precision this minimum ration of albuminoids—or as we may briefly put it,

[1] It is not certain, however, that all the precautions taken have the desired result. You cannot entirely deprive meat of its carbohydrates.

of *albumen*—below which the organism would perish. Voit had found 118 grammes of albumen necessary for the average adult man weighing 70 kilos. This figure is certainly too high. The Japanese doctors, Mori, Tsuboï, and Murato, have shown that a considerable portion of the population of Japan is content with a diet much poorer in nitrogen, and suffers no inconvenience. The Abyssinians, according to Lapicque, ingest, on the average, only 67 grammes of albumen per day. A Scandinavian physiologist, Siven, experimenting on himself, found that he could reduce the ration of albumen necessary to the maintenance and equilibrium of the organism to the lowest figures which have been yet reached—namely, from 35 to 46 grammes a day. These experiments, however, must be confirmed and interpreted. Besides, it is important to point out that the most advantageous ration of albumen requires to be a good deal above the strictly sufficient quantity.

It only remains to refer to several other recent researches. The most important of many are those published by M. Chauveau, on the reciprocal transformation of the immediate principles in the organism according to the conditions of its functioning and the circumstances of its activity. To deal with these researches with as much detail as they deserve, we must study the physiology of muscular contraction and of movement—that is to say, of muscular energetics.

BOOK III.

THE CHARACTERS COMMON TO LIVING BEINGS.

CHAPTER I.

THE DOCTRINE OF VITAL UNITY.

Phenomena common to all living beings—Theory of vital duality—Unity in the formation of immediate principles—Unity in the digestive acts—The common vital fund.

WHEN we ask the various philosophical schools what life is, some show us a chemical retort, and others show us a soul. Whether vitalists or of the mechanical school, these are the adversaries who since philosophy began have vainly contested the possession of the secret of life. We need not concern ourselves with this eternal quarrel. We need not ask Pythagoras, Plato, Aristotle, Hippocrates, Paracelsus, Van Helmont, and Stahl what idea they formed of the vital principle; nor need we probe to the depths the ideas of living nature held by Epicurus, Democritus,

Boerhaave, Willis, and Lamettrie; nor need we apply to the iatromechanicians nor to the chemists. We may do better than that. We may ask nature itself.

Phenomena Common to Living Beings. — Nature shows us an infinite number of beings, animal or vegetable, described in ordinary language as *living beings*. This language implicitly assumes something common to them all, a universal manner of being which belongs to them without distinction, without regard to differences of species, types, or kingdoms. On the other hand, anatomical analysis teaches us that animated beings and plants may be divided into parts ever decreasing in complexity, of which the last and the simplest is the *anatomical element*, the *cell*, the microscopic organic unit which, too, is alive. Common opinion suspects that all these beings, whether entire as in the case of animal and vegetable individuals, or fragmentary as in the case of cellular elements, have the same manner of being, and present the same body of common characteristics which rightly gives them this unmistakable title of living beings. Life then essentially would be this manner of being, common to animals, vegetables, and their elements. To seize in isolation these common, necessary, and permanent features, and then to synthetize them into a whole, will be the really scientific method of defining life, and of explaining its nature.

And here then immediately arises a fundamental question which gives one pause, a question of fact which must be solved before we can go further. Is there really a common manner of being in all these things? Are *animal life*, *vegetable life*, and the life of the elements or *elementary life*, all the same? Is there a sum total of characteristics which may define life in general?

The physiologists, following in the steps of Claude Bernard, respond in the affirmative. They accept as valid and convincing the proof given of this vital community by the illustrious experimentalist. However, there are some rare exceptions to this universal assent. In this concert of approval there is at least one discordant voice, that of M. F. Le Dantec.[1]

[1] M. Le Dantec, of whose philosophical and rigorously systematic mind I have the highest opinion, has laid down a new conception of life, the essential basis of which is this very distinction between elementary life and ordinary life; between the life of the elements or of the beings formed from a single cell, protophytes and protozoa, and the life of ordinary animals and plants, which are multicellular complexes, and for that reason called *metazoa* and *metaphytes*.

Further, in the *elementary life* peculiar to monocellular beings (protozoa and cellular elements), M. Le Dantec distinguishes three manners of being:—The first condition, which is elementary life manifested in all its perfection, cellular health; the second condition is deteriorated elementary life, *cellular disease;* and the third condition, which is *latent life*. I should say at once that in so far as the fundamental distinction of the phenomena of *elementary life* and those of the general life of animals and ordinary plants, metazoa or metaphytes, is concerned, we find it neither justified nor useful. And further, *manifested elementary life*, as M. Le Dantec understands it, would only belong to a small number of *elementary beings*—for the protozoa, starting with the infusoria, are not among the number—and to a still smaller number of *anatomical elements*, since among the vertebrates we recognize as almost the only elements satisfying it, the ovule, and perhaps the leucocyte. Physiologists, therefore, do not agree with M. Le Dantec as to the utility of adding one condition more to those we all admit—namely, manifested animal life and latent life.

The Doctrine of the Vital Duality of Animals and Plants.—There are, therefore, biologists who, in the domain of theory and in virtue of more or less well-founded conceptions or interpretations, separate *elementary life* from other vital forms, and thus break the bond of vital unity proclaimed by Claude Bernard. This monistic doctrine at the outset met with other opponents, and that, too, in the domain of facts. But it triumphed over them and became established. We have to deal with scientists like J. B. Dumas and Boussingault, who drew a dividing line between *animal life* and *vegetable life.*

But let us in a few words recall to the reader this victorious struggle of the monistic doctrine against the dualism of the two kingdoms. If we consider an animal in action, said the champions of vital dualism, we agree that it feels, moves, breathes, digests, and finally, that it destroys by a real operation of chemical analysis the materials afforded to it by its ambient world. It is in these phenomena that are manifested its activity, its life. Now, added the dualists, plants do not feel, do not move, do not breathe, and do not digest. They build up from immediate principles, by an operation of chemical synthesis, the materials they borrow from the soil which bears them, or from the atmosphere which surrounds them. There is, therefore, nothing in common between the representatives of the two kingdoms if we confine ourselves to the examination of the actual phenomena which take place in them. To find a resemblance between the animal and the vegetable, said the dualists, we must set aside what they *do*, for they do different, or even contrary things. We must consider whence they come and what they *become.* Both originate in organisms similar to themselves. They grow, evolve, and generate as they themselves were generated. In other words, while their acts separate plants from animals, their mode of origin and evolution alone bring them together. Such analogies are of no slight importance; but they were neutralized by their dissimilarities, which were exaggerated by the dualistic school.

It is clear that the word *life* would lose all actual significance to those who would reduce it to the faculty of evolution, and who would separate all its real manifestations in animated beings and in plants. If there are two lives, the one animal and the other vegetable, there are no more; or, what comes to the same thing, there is an infinite number of lives which have nothing in common but the name, or at most, the possession of some secondary characteristics. There are as many of them as there are different beings, for each has its own particular evolution. Here the specific is the negation of the general and it destroys it instead of being subordinate to it. The principle of life becomes for each being something as individual as its own evolution. And this, if we think it out, is how the philosophers look at life, and it is the real reason of their disagreement with the physiological school.

Proof of the Monistic Theory.—On the other hand, under the disguise of living forms, the physiologist recognizes the existence of an identical basis. His trained ear marks amid the overcharged instrumentation of the vital work the recognizable undertones of a constant theme. It was the work of Claude Bernard to bring this common basis to light. He shows that plants live as animals do, that they breathe, digest, have sensory reactions, move essentially like animals, destroy and build up in the same manner the immediate chemical principles. For that purpose it was necessary to pass in review, examining them from their foundation and distinguishing the essential from the secondary, the different vital manifestations—digestion, respiration, sensibility, motility, and nutrition. This is what Claude Bernard did in his work *Sur les Phénomènes de la vie communs aux animaux et aux plantes.* We need only to sketch in broad outline the characteristic features of his lengthy demonstration.

Unity in the Formation of Immediate Chemical Principles.—The first and most important of the differences pointed out between the life of animals and that of plants was relative to the formation of immediate principles. On this ground, indeed, vital dualism raised its fortress. The animal kingdom was considered in its totality as the parasite of the vegetable kingdom. To J. B. Dumas, animals, whatever they may be, make neither fat nor any elementary organic matter; they borrow all their foods, whether they be sugars or starches, fats or nitrogenous substances, from the vegetable kingdom. About the year 1843 the researches of the chemists, and of Payen in particular, succeeded in proving the presence, almost constant, of fatty matters in vegetables; and, further, these matters existed there in proportions more than sufficient to explain how the beast which fed upon them was fattened. The chemists attributed to nature as much practical sense as they themselves possessed; and since the hay and the grass of the ration brought fat ready made to the horse, the cow, and the sheep, they declared that the animal organism had nothing whatever to do but to put this food into the tissues, or to arrange for it to pass into the milk. But nature is not so wise and economical as was supposed at the Académie des Sciences. After a memorable debate, in which Dumas, Boussingault, Payen, Liebig, Persoz, Chossat, Milne-Edwards, and Flourens took part, and, later on, Berthelot and Claude Bernard, it was agreed that the animal does not grow fat from the fatty food which is supplied it, and that it makes its own fat just as the vegetable does, but in another manner. In the same way sugar, the normal constituent substance necessary for the nutrition of animals and plants, instead of being a vegetable product passing by alimentation from the herbivorous animals and thence to the carnivorous, is manufactured by the animal itself. Generally speaking, immediate principles have an equal claim to existence in the two kingdoms. Both form and destroy the substances indispensable to life.

Here, then, one of the barriers between animal life and vegetable life is overthrown and destroyed.

Unity of Digestive Acts in Animals and Plants.—Similarly, another barrier falls if we show that digestion, long considered the exclusive function of animals, and, in particular, of the higher animals, is in reality universal.

Cuvier pointed out the absence of a digestive apparatus as a very general and distinctive characteristic of plants. But the absence of a digestive apparatus does not necessarily imply the absence of digestion. The essential act of digestion is independent of the infinite variety of the organs, just as a reaction is independent of the form of the vessel in which it takes place. It is, in fact, a chemical transformation of an alimentary substance. This transformation may be realized outside the organism, *in vitro*, just as it can in the living being without masticating organs, without an intestinal apparatus, without glands, in a vessel placed in a stove, simply by means of a few soluble ferments—pepsine, trypsine, amylolytic diastases.

All alimentary substances, whether taken from without or borrowed from the reserves accumulated in the internal stores of the organism, must undergo preparation. This preparation is digestion. Digestion is the prologue of nutrition. It is over when the reparative substance, whether food or reserve-stuff, is brought into a state enabling it to pass into the blood, and to be utilized by the organism.

The Identity of Categories of Foods in the Two Kingdoms.—Now the alimentary substances are the same in the two kingdoms, and so is their digestive preparation. Alimentary materials are of four kinds: albuminoid, starchy, fatty, and sugary substances. The animal takes them from without (food properly so-called), or from within (reserve-stuff). Man obtains starch, for instance, from different farinaceous dishes. It may, however, equally well be borrowed from the reserve of flour that we carry within us in our liver, which is a veritable granary, full of floury substance, glycogen. And so it is with vegetables. The potato has its store of flour in its tuber just as the animal has in its liver. The grain which is about to germinate has it in reserve-stuff in its cotyledons, or in its albumen. The bud which is about to develop into a tree or a flower carries it at its base.

The same conclusions are true for another class of substances, the sugars. They may be a food taken from without, or a reserve deposited in the tissues. The animal takes from without, in fruits for instance, the ordinary sugar which pleases its taste. Beetroot, when flowering and fructifying, draws this substance from its roots in which stores have been amassed. The sugar cane when running to seed takes the sugar from the stores which it possesses in its cane. Brewer's yeast, the *saccharomyces cerevisiæ*, the agent of alcoholic fermentation, finds this same substance in the sugary juices favourable to its development.

In the same way, identically fatty substances, either in the form of food or of reserve-stuff, serve for nutrition to animals and vegetables; and that is again true of the substances of the fourth class, albuminoids, identical in the two kingdoms, foods or reserve-stuff, equally utilizable in both after digestion.

Identity of the Digestive Agents and Mechanisms in Plants and Animals.—Now, the results of contemporary research have been to establish a surprising resemblance in the modifications experienced by these foods, or reserve stuffs, in animals and plants; and even resemblances in the agents which realize them, and in the mechanisms by which they are performed. There is a real unity. The flour accumulated in the tuber of the potato is liquefied and digested on the appearance of the buds or of the flower, just as the starch of the liver or the alimentary flour is digested by the animal. The fatty matter which is stored up in the oleaginous grain is digested at the moment of germination, just as the fat during a meal is digested in the animal's intestine. As the beetroot begins to run to seed, the root gives up part of its store of sugar, and this reserve stuff is distributed throughout the stalk after having been digested, exactly as would have been the case in the digestive canal of man.

Vegetables, then, really digest. The four classes of substances mentioned above are really digested in order to pass from their actual form, a form unsuitable for interstitial exchanges, to another form suitable for nutrition. As there are four kinds of foods, so there are four kinds of digestions, four kinds of ferment-producing agents—amylolytic,[1] proteolytic,[2] sacchar-ine, and lipasic[3] diastases, identical in the animal and the plant. Identity of ferments implies identity of digestions. Going down to the very basis of things, the digestive act is nothing but the action of this ferment. This is the crux of the whole question. All else is only difference in scene, varying in the means of execution and in the accessories. The difference arises from the stage on which it takes place, but the piece which is being played is the same, and the actors are the same, and so is the action of the play.

This identity between animal and vegetable life is found in the phenomena of respiration and of motility. The limits of this book do not allow of our entering into the details of facts. Besides, the facts are well known, and may be found in any treatise on general physiology. This science, therefore, enables us to perceive the imposing unity of life in its essential manifestations.

[1] Amylolytic ferments change starch and glycogen (*amyloses*) into sugar.—TR.

[2] Proteolytic ferments change proteids into peptones and proteoses.—TR.

[3] The enzyme known as lipase splits the fat or oil in germinating seeds into a fatty acid and glycerine.—TR.

The community of the phenomena of vitality in animals and plants being thus placed beyond a doubt, we must now discover the reason why. This reason is to be found in their anatomical and in their chemical unity. The fundamental phenomena are common because the composition is common, and because the universal anatomical basis, the cell, possesses in all cases a sum total of identical properties.

If we appeal to physiology for the characteristics common to living beings, it will generally give us the following:—A structure or organization; a certain chemical composition which is that of *living matter;* a specific form; an evolution which in the earliest stage occasions the being to grow and develop until it is divided, and which in the highest stage includes one or more evolutive cycles with growth, the adult stage, senility, and death; a property of increase or nutrition, with its consequence—namely, a relation of material exchanges with the ambient medium;—and finally, a property of reproduction. It is important to pass them rapidly in review.

CHAPTER II.

MORPHOLOGICAL UNITY OF LIVING BEINGS.

THE first characteristic of the living beings is *organization.* By that we mean that they have a structure; that they are complex bodies formed of smaller aliquot parts and grouped according to a certain disposition. The most simple elementary being is not yet homogeneous. It is heterogeneous. It is organized. The least complex protoplasms, that of bacteria, for example, still possess a physical structure; Kunstler distinguishes in them two non-miscible substances, presenting an alveolar organization. Thus animals and plants present an organization, and it is sensibly constant from one end to the other of the scale of beings. There is a *morphological unity.*

§ 1. THE CELLULAR THEORY. FIRST PERIOD: DIVISION OF THE ORGANISM INTO CELLS.

Cellular Theory. First Period.—Morphological unity results from the existence of a universal anatomical basis, the *cell.* The cellular theory sums up the teaching of general anatomy or histology.

At the beginning of the nineteenth century anatomy was following a routine dating from ancient times. It divided animal and vegetable machines into units in descending order, first into different forms of apparatus (circulatory, respiratory, digestive, etc.); then the apparatus into organs examined one by one, figuring and describing each of them from every point of view with scrupulous accuracy and untiring patience. If we think of the duration of these researches—the *Iliad*, as Malgaigne says, already containing the elements of a very fine regional anatomy—and especially of the powerful impulse they received in the seventeenth and eighteenth centuries, we shall understand the illusion of those who, in the days of X. Bichat, could fancy that the task of anatomy was almost ended.

As a matter of fact this task was barely begun, for nothing was known of the intimate structure of the organs. X. Bichat accomplished a revolution when he decomposed the living body into tissues. His successors, advancing a step in the analysis, dissociated the tissues into elements. These elements, which one would have thought were infinitely varied, were reduced in their turn to one common *prototype*, the cell.

The living body, disaggregated by the histologist, resolves under the microscope into a dust, every grain of which is a cell. A cell is an anatomical element the constitution of which is the same from one part to the other of the same being, and from one being to another; and its dimensions, which are sensibly constant throughout the whole of the living world, have an average diameter of several thousandths of a millimetre—*i.e.*, of several *microns*. This element, the cell, is a real organ. It is smaller, no doubt, than those described by the ancient anatomists, but it is not less complex. Its complexity is only revealed later. It is an organic unit. Its form varies from one element to another. Its substance is a semi-fluid mass, a mixture of different albuminoids. In the mean value of its dimensions, so carefully measured—*exceptis excipiendis*—we have a condition the significance of which has not yet been discovered, but which may be of great value in the explanation of its peculiar activities.

Such is the result to which have converged the researches of the biologists who have examined plants or the lower animals, as well as of the anatomists who have been more especially occupied with the vertebrates and with man. All their researches have brought them to the same conclusion—the cellular theory. Either living beings are composed of a single cell—as is the case with the microscopic animals called *protozoa*, and the microscopic vegetables called *protophytes*—or, they are cellular complexes, *metazoa* or *metaphytes*—that is to say, associations of these microscopic organic units which are called cells.

The Law of the Composition of Organisms.—The law of the composition of organisms was discovered in 1838 by Schleiden and Schwann. From that time up to 1875 it may be said that micrographers have spent their time in examining every organ and every tissue, muscular, glandular, conjunctive, nervous, etc., and in showing that in spite of their varieties of aspect and form, of the complexity of structures due to cohesion and fusion, they all resolve into the common element, the cell. Contemporary anatomists, Koelliker, Max Schultze, and Ranvier, have thus established the generality of the cellular constitution, while zoologists and botanists confirm the same law for all animals and vegetables, and exhibit them all as either unicellular or multicellular.

The Cellular Origin of Complex Beings.—At the same time embryogenic researches showed that all beings spring from a corpuscle of the same type. Going back in the history of their development to the most remote period, we find a cell of very constant constitution—namely, the *ovule.* This truth may be expressed by changing a word in Harvey's celebrated aphorism—*omne vivum ex ovo;* we now say *omne vivum e cellula.* The myriads of differentiated anatomical elements whose association forms complex beings are the posterity of a cell, of the *primordial ovule,* unless they are the posterity of another equivalent cell. The second task of histology in the latter half of the nineteenth century consisted in following up the filiation of each anatomical element from the cell-egg to its state of complete development.

The whole cellular theory is contained in the two following statements, which establish the morphological unity of living beings:—*Everything is a cell, everything comes from an initial cell;* the cell being defined as a mass of substance, protoplasm or protoplasms, of an average diameter of a few microns.

§ 2. The Second Period: the Division of the Cell.

Second Period: Constitution of the Cell.—This was, however, only the first phase in the analytical study of the living being. A second period began in 1873 with the researches of Strassburger, Bütschli, Flemming, Kuppfer, Fromann, Heitzmann, Balbiani, Guignard, Kunstler, etc. These observers in their turn submitted this anatomical, this infinitely small cellular microcosm, to the same penetrating dissection their predecessors had applied to the whole organism. They brought us down one degree lower into the abyss of the infinitely small. And as Pascal, losing himself in these wonders of the imperceptible, saw in the body of the mite which is only a point, "parts incomparably smaller, legs with joints, veins in the legs, blood in the veins, humours in the blood, drops in the humours, vapours in these drops," so contemporary biologists have shown in the epitome of organism called a cell, an edifice which itself is marvellously complex.

The Cytoplasm.—The observers named above revealed to us the extreme complexity of this organic unit. Their researches have shown us the structure of the two parts of which it is composed—the cellular protoplasm and the nucleus. They have determined the part played by each in genetic multiplication. They have shown that the protoplasm which forms the body of the cell is not homogeneous, as was at first supposed. The idea which was mooted later, that this protoplasm was formed, to use Sachs' words, of a kind of "protoplasmic mud,"—*i.e.*, of a dust consisting of grains and granules connected by a liquid,—is no longer accurate. There is a much simpler view of the case. According to Leydig and his pupils, we must compare the protoplasm to a sponge in the meshes of which is lodged a fluid, transparent, hyaline substance, a kind of cellular juice, hyaloplasm. From the chemical point of view this cellular juice is a mixture of very different materials, albumens, globulins, carbohydrates, and fats, elaborated by the cell itself. It is a product of vital activity; it is not yet the seat of this activity. The living matter has taken refuge in the spongy tissue itself, in the *spongioplasm*.

According to other histologists, the comparison of protoplasm to a spongy mass does not give the most exact idea, and, in particular, it does not furnish the most general idea. It would be far better to say that the protoplasm possesses the structure of foam or lather. As was seen by Kunstler in 1880, a comparison with some familiar objects gives the best idea. Nothing could be more like protoplasm physically than the culinary preparation known as *sauce mayonnaise*, made with the aid of oil and a liquid with which oil does not mix. Emulsions of this kind were made artificially by Bütschli. He noted that these preparations mimicked all the aspects of cellular protoplasm. Thus, in the living cell there is a mixture of two liquids, non-miscible and of unequal fluidity. This mixture gives rise to the formation of little cells. The more consistent substance forms their supporting framework (Leydig's spongioplasm), while the other, which is more fluid, fills its interior (hyaloplasm).

However that may be, whether the primitive organization of the cellular protoplasm be that of a sponge, as is asserted by Leydig, or that of a *sauce mayonnaise*, as is claimed by Bütschli and Kunstler, the complexity does not rest there. Further recourse must be made to analysis. Just as the tissue of a sponge, when torn, shows the fibres which constitute it, so the spongioplasm, the parietal substance, is exhibited as formed of a tangle of fibrils, or better still, of filaments or ribbons (in Greek, *mitome*), which are called *chromaticifilaments*, because they are deeply stained when the cell is plunged into aniline dye. In each of these filaments, the substance of which is called chromatin, the devices of microscopic examination enable us to discover a series of granulations like beads on a string, the *microsomes* or bioblasts, connected one with the other by a sort of cement, Schwartz's *linin*, which is a kind of nuclein.

And let us add, to complete this summary of the constitution of cellular protoplasm, that it presents, at any rate at a certain moment, a remarkable organ, the *centrosome*, which plays an important part in cellular division. Its pre-existence is not certain. Some writers make it issue from the nucleus. At the moment of cellular division it appears like a compressed mass of granulations, which may be deeply stained. Around it is seen a clear unstainable zone, called the attraction-sphere; and finally, beyond this is a crown of striæ, which diverge like the rays of a halo—*i.e.*, the *aster*. In conclusion, there are yet in the cellular body three kinds of non-essential bodies: the vacuoles, the leucites, and various inclusions. The *vacuoles* are cavities, some inert, some contractile; the *leucites* are organs for the manufacture of particular substances; the *inclusions* are the manufactured products, or wastes.

The Nucleus.—Every cell capable of living, growing, and multiplying, possesses a *nucleus* of constitution very analogous to the cellular mass which surrounds it. The anatomical elements in which no nucleus is found, such as the red globules of blood in adult mammals, are bodies which are certain, sooner or later, to disappear. There is therefore no real cell without a nucleus, any more than there is a nucleus without a cell. The exceptions to this law are only apparent. Histologists have examined them one by one, and have shown their purely specious character. We may therefore lay aside, subject to possible appeal from this decision, organisms such as Haeckel's *monera* and the problem of finding out if bacteria really have a nucleus. The very great, if not the absolute generality of the nuclear body, must be admitted.

It hence follows that there is a nuclear protoplasm and a nuclear juice, just as we have seen that there is a protoplasm and a cellular juice. What was just said of the one may now be repeated of the other, and perhaps with even more emphasis. The nuclear protoplasm is a filamentary mass sometimes formed of a single mitome or cord, folded over on itself and capable of being unrolled. The mitome in its turn is a string of microsomes united by the cement of the linin. These are the same constituent elements as before, and the language of science distinguishes them one from the other by a prefix to their name of the words *cyto* or *karyo*, which in Greek signify cell and nucleus, according as they belong to one or the other of these organs. These are mere matters of nomenclature, but we know that in the descriptive sciences such matters are not of minor importance.

We have just indicated that in a state of repose,—that is to say, under ordinary conditions,—the structure of a nucleus reproduces clearly the structure of the cellular protoplasm which surrounds it. The nuclear essence is best separated from the spongioplasm. It takes more clearly the form of a filamentary thread, and the filaments themselves (mitome) show very thick chromatic granulations, or microsomes, connected by the linin.

At the moment of reproduction of the cell these granulations blend into a stainable sheath which surrounds the filaments, and the latter dispose themselves so as to form a single thread. This chromatic filament, which has now become a single thread, is shortened as it thickens (*spireme*); it is then cut into segments, twelve or twenty-four in the case of animals and a larger number in the case of plants. These are *chromosomes*, or *nuclear segments*, or *chromatic* loops. Their part is a very important one. They are constant in number and permanent during the whole of the life of the cell. Let us add that the nucleus still contains accessory elements (nucleoli).

The Rôle of the Nucleus.—Experiment has shown that the nucleus presides over the nutrition, the growth, and the conservation of the cell. If, following the example of Balbiani, Gruber, Nussbaum, and W. Roux of Leipzig, we cut into two a cell without injuring the nucleus, the fragment which is denuded of the nucleus continues to perform its functions for some time in the ordinary manner, and in some measure in virtue of its former impulse. It then declines and dies. On the contrary, the fragment provided with the nucleus repairs its wound, is reconstituted and continues to live. Thus the nucleus takes a very remarkable part in the reproduction of the cell, but it is still a matter of uncertainty whether its rôle is here subordinated to that of the cellular body, or if it is pre-eminent. However that may be, it follows from this experiment that the nucleus presents all the characteristics of a vigorous vitality, and that it is in its protoplasm that the chemists should be able to find the compounds, the special albuminoids, which, *par excellence*, form living matter.

§ 3. THE PHYSICAL CONSTITUTION OF LIVING MATTER. THE MICELLAR THEORY.

Physical Constitution of Living Matter.—Microscopic examination does not take us much farther. The microscope, with the strongest magnification of which it is capable at present, shows us nothing beyond these links of aligned microsomes forming the species of protoplasmic thread or mitome, whose cellular body is a confused tangle or a very tangled ball. It is not probable that direct sight can penetrate much farther than this. No doubt the microscope, which has been so vastly improved, is capable of still further improvement. But these improvements are not indefinite. We have already reached a linear magnification of 2000, and theory tells us that a magnification of 4000 is the limit which cannot be passed. The penetrating power of the instrument is therefore near its culminating point. It has already given almost all that we have a right to expect from it.

We must, however, penetrate beyond this microscopic structure at which the sense of sight has been arrested. How is this to be done? When observation is arrested, hypothesis takes its place. Here there are two kinds of hypotheses, the one purely anatomical, the other physical. Anatomically, beyond the visible microsomes there have been imagined invisible hyper-microscopic corpuscles, the plastidules of Haeckel, the idioblasts of Hertwig, the pangenes of de Vries, the plasomes of Wiesner, the gemmules of Darwin, and the biophores of Weismann.

Biologists who have not got all that they hoped from microscopic structure are therefore thrown back on hyper-microscopic structure.

It is very remarkable that all this profound knowledge of structure has been so sterile from the point of view of the knowledge of cellular functional activity. All that is known of the life of the cell has been revealed by experiment. Nothing has resulted from microscopic observation but ideas as to configuration. When it is a question of giving or imagining an explanation of vital facts, of heredity, etc., biologists unable to supply anything beyond the details of structure revealed by anatomy have had recourse to hypothetical elements, gemmules, pangenes, biophores, and different kinds of determinants.

Anatomy never has explained and never will explain anything. "Happy physicists!" wrote Loeb, "in never having known the method of research by sections and stainings! What would have happened if by chance a steam engine had fallen into the hands of a histological physicist? How many thousands of sections differently stained and unstained, how many drawings, how many figures, would have been produced before they knew for certain that the machine is an engine, and that it is used for transforming heat into motion!"

The study of physical properties, continued on rational hypotheses, has also thrown some light on the possible constitution of living matter. The gap between microscopical structure and molecular or chemical structure has thus been filled.

The consideration of the properties of *turgescence* and of *swelling*, which very generally belong to organized tissues, and therefore to the organic substance of protoplasm, has enabled us to obtain some idea of its ultra-microscopic constitution. If we wet a piece of sugar or a morsel of salt, before they are dissolved they absorb and imbibe the water without sensibly increasing their volume. It is quite otherwise with a tissue (*i.e.*, with a protoplasm) when weakened in water as a preliminary. The tissue, plunged into the liquid, absorbs it, swells, and often grows considerably. And this water does not lodge in the gaps, in pre-existing lacunar spaces, for organic matter presents no gaps of this kind. It does not resemble a porous mass with capillary canals, such as sandstone, tempered mortar, clay, or refined sugar. The molecules of water interpose between and separate the organic molecules, thus increasing by a sort of intussusception the intervals separating the one from the other—molecular intervals escaping the senses, as do the molecules themselves because they are of the same order of magnitude.

Micellar Theory. — While pondering over this phenomenon, an eminent physiologist, Nägeli, was led in 1877 to propose his *micellar theory*. Micellæ are groups of molecules in the sense in which physicists and chemists use the word. They are molecular structures with a configuration. They rapidly absorb water and are capable of fixing a more or less thick and adherent layer of it to their surface. In a word, they are aggregates of organic matter and water.

There is therefore every reason for believing that the *microsomes* of spongy protoplasm, the physical support or basis of cellular life, are *groups of micellæ* formed of albuminoid substances and water. These clustered forms, these micellæ, are not absolutely peculiar to organized matter. Pfeffer, the learned botanist, has pointed them out under another name, *tagmata*, in the membranes of chemical precipitates.

Beyond this limit analysis finds nothing but the chemical molecule and the atom. So that if we wish to reconstruct the hierarchy of the materials of constitution of the protoplasm in order of ascending complexity, we shall find at the foundation the atom or atoms of simple bodies. They are principally carbon, hydrogen, oxygen, nitrogen, the elements of all organic compounds, to which may be added sulphur and phosphorus. At the head we have the albuminoid molecule, or the albuminoid molecules, aggregates of the preceding atoms. In the third stage the micellæ or tagmata, aggregates of albuminoids and water, are still too small to be observed by the senses. They unite in their turn to form the microsomes, the first elements visible to the microscope. The microsomes, cemented by linin, form the filaments or links which are called mitomes. The living protoplasm is therefore nothing but a chain, or tangled skein, or a spongy skeleton formed by its filaments.

Such is the typical constitution of living matter according to microscopic observation, supplemented by a perfectly reasonable hypothesis, which is, so to speak, only a translation of one of its most evident physical properties. This relatively simple scheme has become a complex scheme in the hands of later biologists. On the micellar hypothesis, which seems almost inevitable in its character, new hypotheses have been grafted, merely for the sake of convenience. Hence, we are led farther and farther from the real truth, and this is why, in order to explain the phenomena of heredity, we find ourselves compelled to intercalate hypothetical elements between micellæ and the microsome in the higher hierarchy quoted above—gemmules, pangenes, plasomes, which are only mental pictures or simple images to represent them.

§ 4. The Individuality of Complex Beings. Law of the Constitution of Organisms.

Individuality of Complex Beings.—From the cellular doctrine follows a remarkably suggestive conception of living beings. The metazoa and the metaphytes—that is to say, the multicellular living beings which may be seen with the eyes and do not require the microscope to reveal them—are an assemblage of anatomical elements and the posterity of a cell. The animal or the plant, instead of being an individual unity, is a "multitude," a term which is used by Goëthe himself when pondering, in 1807, over the doctrine taught by Bichat; or, according to the equally correct expression of Hegel, it is a "nation"; it springs from a common cellular ancestor, just as the Jewish people sprang from the loins of Abraham.

We now picture to ourselves the complex living being, animal or plant, with its configuration which distinguishes it from every other being, just as a populous city is distinguished by a thousand characteristics from its neighbour. The elements of this city are independent and autonomous for the same reason as the anatomical elements of the organism. Both have in themselves the means of life, which they neither borrow nor take from their neighbours nor from the whole. All these inhabitants live in the same way, are nourished and breathe in the same manner, all possessing the same general faculties, those of man; but each has besides, his profession, his trade, his aptitudes, his talents, by which he contributes to social life, and by which in his turn he depends on it. Professional men, the mason, the baker, the butcher, the manufacturer, the artist, carry out different tasks and furnish different products, the more varied, the more numerous and the more differentiated, in proportion as the social state has reached a higher degree of perfection. The living being, animal or plant, is a city of this kind.

Law of the Constitution of Organisms.—Such is the complex animal. It is organized like the city. But the higher law of this city is that the conditions of the elementary or individual life of all the anatomical citizens are respected, the conditions being the same for all. Food, air, and light must be brought everywhere to each sedentary element; the waste must be carried off in discharges which will free the whole from the inconvenience or the danger of such debris; and that is why we have the different forms of apparatus in the circulatory, respiratory, and excretory economy. The organization of the whole is therefore dominated by the necessities of cellular life. This is expressed in *the law of the constitution of organisms* formulated by Claude Bernard. The organic edifice is made up of apparatus and organs, which furnish to each anatomical element the necessary conditions and materials for the maintenance of life and the exercise of its activity. We now understand what is the life, and at the same time what is the death, of a complex being. The life of the complex animal, of the metazoon, is of two degrees; at the foundation, the activity proper to each cell, *elementary life*, cellular life; above, the forms of activity resulting from the association of the cells, *the life of the whole*, the sum or rather the complex of elementary partial lives. There is a solidarity between them produced by the nervous system, by the community of the general circulatory, respiratory apparatus, etc., and by the free communication and mixture of the liquids which constitute the media of culture for each cell. We shall have an opportunity of recurring to current ideas as to the morphological constitution of organisms.

CHAPTER III.

THE CHEMICAL UNITY OF LIVING BEINGS.

The varieties and essential unity of the protoplasm—Its affinity for oxygen—The chemical composition of protoplasm—Its characteristic substances.—§ 1. The different categories of albuminoid substances—Nucleo-proteids—Albumins and histones—Nucleins.—§ 2. Constitution of nucleins.—§ 3. Constitution of histones and albumins—Schultzenberger's analysis of albumin—Kossol's analysis—The hexonic nucleus.

THE chemical unity of living beings corresponds to their morphological unity.

The Varieties and Essential Unity of the Protoplasm.—One essential feature of the living being is that it is composed of matter peculiar to it, which is called *living matter*, or *protoplasm*. But this is a somewhat incorrect way of expressing the facts. There is no unique living matter, no single protoplasm; their number is infinite, there are as many as there are distinct individuals. However like one man may be to another, we are compelled to admit that they differ according to the substance of which they are constituted. That of the first offers a certain characteristic personal to the first, and found in all his anatomical elements; similarly for the second. With Le Dantec we shall say that the chemical substance of Primus is not only of the substance of man, but in all parts of his body and in all his constituent cells it is the exclusive substance of Primus; and, in the same way, the living matter of another individual Secundus will carry everywhere his personal impress, which differs from that of Primus.

But it is none the less true that this absolute specificity is based with certainty only on differences which from the chemical point of view are exceedingly slight. All these protoplasms have a very analogous composition. And, if we regard as negligible the smallest individual, specific, generic, or ordinal variations we may then speak in a general manner of *protoplasm* or *living matter.*

Experiment shows us, in fact, that the real living substance—apart from the products it manufactures and can retain or reject—is in every cell tolerably similar to itself. The fundamental chemical resemblance of all protoplasms is certain, and thus we may speak of their typical composition. We may sum up the work of physiological chemistry for the last three quarters of a century by affirming that it has established the chemical unity of all living beings—that is to say, a very notable analogy in the composition of their protoplasm.

This living matter is essentially a mixture of the proteid or albuminoid substances, to which may be added other categories of immediate principles, such as carbohydrates and fatty matters. But the latter are of secondary importance. The essential element is the proteid substance. The most skilful chemists have tried for more than half a century to discover its composition. Only during the last few years —thanks to the researches of Kossel, the German chemist, following on those of Schultzenberger and Miescher—we are beginning to know the outer walls or the framework of the albuminoid molecule; in other words, its chemical nucleus.

Physical Characters of Protoplasm.—About 1860 Ch. Robin thought that he had defined living matter sufficiently—or, at least, as perfectly as could be expected at that time—by attributing to it three physical characteristics. They were:—Absence of homogeneity, molecular symmetry, and the association of three orders of immediate principles—albuminoids, carbohydrates and fats. These characteristics assist, but do not suffice, to define the organization.

No doubt the characteristics must be completed by the addition of a certain number of more subtle physical features.

One of them refers to the structure of protoplasm as revealed by the microscope. Throughout the whole of the living kingdom, from the bacteria studied by Kunstler and Busquet to the most complicated protozoa, protoplasmic matter presents the same constitution, and in consequence, this structure of the protoplasm must be considered as one of its distinctive characters. It is not homogeneous; it is not the last term of the visible organization: it is itself organized. Experiment shows that it does not resist breaking up or crushing. Mutilations cause it to lose its properties. As for the kind of structure that it presents, it may be expressed by saying that it is that of a foamy emulsion.

We saw above that our knowledge as to the physical condition of protoplasm has been completed by the theories of Bütschli's micellæ or Pfeffer's tagmata.

Properties of the Protoplasm. Its Affinity for Oxygen.—From the chemical point of view, living matter presents a very remarkable property—namely, a great affinity for oxygen. It absorbs it so greedily that the gas cannot remain in a free state in its neighbourhood. Living protoplasm, therefore, exercises a reducing power. But it does not absorb oxygen in this way for its own advantage; oxygen is not absorbed, as was supposed thirty years ago, to supply fuel wherewith to burn the protoplasm. The products are not those of its oxidation, of its own disintegration. They are the products of combustion of the reserve-stuff which is incorporated in it. These substances have been supplied to it from without, like the oxygen itself, with the blood. This was proved by G. Pflüger in 1872 to 1876. The protoplasm is only the focus, the scene, or the factor of combustion. It is not its victim, it does not itself furnish the fuel. It works like the chemist, who obtains a reaction with the substances that are given to him.

As for the reducing power of protoplasm, A. Gautier in 1881 and Ehrlich in 1890 have given fresh proofs. A. Gautier in particular has insisted that the phenomena of combustion take place, so to speak, outside the cell, and at the expense of the products which surround it; while on the contrary the really active and living parts of the nucleus and of the cellular body, work protected by the oxygen, as in the case of anaerobic microbes.

This result is of great importance. Burdon Sanderson, the late learned professor of physiology at the University of Oxford, has not hesitated to compare it to the discovery of respiratory combustion by Lavoisier. There is no doubt some exaggeration in the comparison; but there is, on the other hand, no less exaggeration in supposing that it is not of great importance. We may no longer in these days speak without reservation of the vital vortex of Cuvier, and of the incessant twofold movement of assimilation and dissimilation which is ever destroying living matter and building it up again. In reality, the living protoplasm varies very little; it only undergoes oscillations of very slight extent; it is the materials, the reserve stuff on which it operates, which are subject to continual transformations.

Chemical Composition of Protoplasm.—One of the the three characters attributed by Ch. Robin to living matter was its chemical composition, of which little was known in his time. He insisted on the constant presence in the living elements of three orders of immediate principles—proteid substances, carbohydrates, and fatty bodies. In reality the proteid substances, or albuminoids, alone are characteristic. The two other groups, carbohydrates and fatty bodies, are rather the signs and the products of the vital activity, than constituents of the matter on which it is exercised.

It is therefore on the knowledge of the proteid substances that all the sagacity of biological chemists has been exercised. Their efforts for thirty years, and particularly in the last few years, have not been barren; they enable us to give a first rough sketch of the constitution of these substances.

§ 1. THE CHARACTERISTIC SUBSTANCES OF THE PROTOPLASM. THE NUCLEO-PROTEIDS.

The Different Categories of Albuminoid Substances.—Albuminoid or proteid substances are extremely complex compounds, much more so than any of those which are being constantly studied by the chemist. They also are to be found in great variety. It has been difficult to separate them one from the other, to characterize them rigorously, or, in other words, to classify them. However, it has been done now, and we distinguish three classes which are differentiated at once from the physiological and from the chemical points of view. The first comprises the complete or typical albuminoids. They are the *proteids* or *nucleo-albuminoids.* They are to be found in the most active and most living parts of the protoplasm, and therefore in the spongioplasm of the cell and around the nucleus. The second group is formed of *albumins* and *globulins*, compounds already simpler, fragments derived from the destruction of the preceding, into which they enter as constituent elements. In the isolated state they do not belong to the really living protoplasm; they exist in the cellular juice, in the interstitial and circulating liquids in the blood and in the lymph. The third category comprises real but incomplete albuminoids. They are to be found in the portions of the economy which have a specialized or attenuated life, and are destined to serve as a support to the more active elements—*i.e.*, they contribute to the building up of the bony, cartilaginous, conjunctive, elastic tissues. They are called *albumoids.* It is naturally the first group, that of the proteids—*i.e.*, of the complete and characteristic compounds of the living substance—upon which the attention of the physiologists must be fixed. It is only quite recently that the clear definition of these substances has been given, and proteid compounds detected in the confused mass.

The Nucleo-proteids.—This progress in the characterization and specification of the proteids required in the first place a knowledge of two particular compounds, the *nucleins* and the *histones.* This did not become possible until after the researches of Miescher and Kossel on the nucleins, which went on from 1874 to 1892, and those of Lilienfeld and d'Yvor Bang on the histones, from 1893 to 1899. The complete albuminoids are constituted by the combination of two kinds of substances—albumins or histones on the one hand, and nucleins on the other. By combining solutions of albumins or histones with solutions of nuclein, the synthesis of the proteid is effected. The study of the properties and characteristics of these nucleo-albumins and nucleo-histones is going on at the present moment. It is being carried out with much method and with wonderful patience by the German school.

All the proteids contain phosphorus in addition to the five chemical elements, carbon, oxygen, hydrogen, nitrogen, and sulphur, which are common to the other albuminoids. Another interesting feature in their history is that the action of the gastric juice divides them into their two constituents:—the nuclein, which is deposited and resists the destructive action of the digestive liquid, and the albumin or histone, which on the contrary experiences this action with the usual consequences. Thus the gastric juice furnishes a process which is very simple and very convenient in the analysis of the proteids.

Localization of the Nucleo-Proteids.—What we said before as to the important physiological rôle of the cellular nucleus may arouse the expectation that in it will be found the living matter which is chemically the most differentiated, the albuminoids of highest rank—*i.e.*, the nucleo-proteids and their constituents. Not that they would not be found in the protoplasm of the rest of the cell, but there is certainly a risk that they would be less concentrated there and more blended with accessory products; they are there connected with much more secondary vital functions. This conclusion inspired the early researches of Professor Miescher, of Basle, in 1874, and, twenty years later, those of Professor Kossel, one of the most eminent physiological chemists in Germany.

In fact, these compounds have been found in all tissues which are rich in cellular elements with well-developed nuclei. The white globules of the blood furnished to Lilienfeld the first nucleo-histone ever isolated. The red globules themselves, when they possess a nucleus, which is the case in birds and reptiles as well as in the embryo of mammals, contain a nucleo-proteid which was easily isolated by Plosz and Kossel. Hammarsten, the Swedish chemist, who has acquired a great reputation from his researches in other domains of biological chemistry, prepared the nucleo-proteids of the pancreas in 1893. They have been obtained from the liver, from the thyroid gland (Ostwald), from brewers' yeast (Kossel), from mushrooms, and from barley (Petit). They have been detected in starchy bodies and in bacteria (Galeotti).

§ 2. CONSTITUTION OF NUCLEINS.

Constitution of Nucleins.—Our path is already marked out if we wish to penetrate farther into the constitution of these proteids, which are the immediate principles highest in complexity among those which form the living protoplasm. We must analyze the two components, the albumins and the histones on the one hand, and the nucleins on the other. As for the nucleins, this has already been done, or very nearly so.

Kossel, in fact, decomposed the nuclein by a series of very carefully arranged operations, and has reduced it step by step to its crystallizable organic radicals. At each stage that we descend in the scale of simplification a body appears which is more acid and more rich in phosphorus. At the third stage we come to phosphoric acid itself. The first operation divides the nuclein into two substances: the new albumin and nucleinic acid. After separating these elements they can be reunited: a solution of albumin with a solution of nucleinic acid reconstitutes the nuclein. A second operation separates the nucleinic acid in its turn into three parts. One is a body of the nature of the sugars—*i.e.*, a carbohydrate. The appearance of a sugar in this portion of the molecule of nucleinic acid is an interesting fact and fertile in results. The second part is constituted by a mixture of nitrogenous bodies, well known in organic chemistry under the name of *xanthic bases* (xanthin, hypoxanthin, guanin, and adenin). The third part is a very acid body and full of phosphorus—thymic acid. If in a third and last operation the thymic acid is analyzed, it is finally separated into phosphoric acid and into thymene, a crystallizable base, and thus we are brought back to the physical world, for all these bodies incontestably belong to it.

§ 3. The Constitution of Histones and Albumins.

Constitution of Histones.—But we are only half-way through our task. We are acquainted in its origin with one of the genealogical branches of the proteid, the nucleinic branch. We must also learn something of the other branch, the albumin or histone branch. But on this side the problem assumes a character of difficulty and complexity which is admirably adapted to discourage the most untiring patience.

The analysis of albumin for a long time baulked the chemist. "Here," said Danilewsky, "we come to a closed door which resists all our efforts." We know how vastly interesting what is taking place on the other side must be, but we cannot get there. We get a mere glimpse through the cracks or chinks which we have been able to make.

This analysis of albuminous matter at first requires great precautions. The chemist finds himself in the presence of architecture of a very subtle kind. The molecule of albumin is a complex edifice which has used up several thousand atoms. To perceive the plan and structure, it must be dismantled and separated into parts which are neither too large nor too small. Such careful demolition is difficult. Processes too rough or too violent will reduce the whole to the tiniest of fragments. It is a statue which may be reduced to dust, instead of being separated into recognizable fragments, easily fitted in place along their fractured faces.

Analysis of Albumin by Schützenberger.—Schützenberger, a chemist of great merit, attempted (about

1875) this thankless task. Others before him had experimented in various ways. Two Austrian scientists, Hlasitwetz and Habermann, in 1873, and a little later Drechsel in 1892, had used concentrated hydrochloric acid to break down albumin. They also employed bromine for the same purpose. More recently Fuerth had used nitric acid with a similar object. Schützenberger tried another way. The battering ram which he used against the edifice of albumin was a concentrated alkali, baryta. He warmed the white of an egg with barium hydrate in a closed vessel at a temperature of 200°. The albumin of egg then divides into a certain number of simpler groups. The difficulty is to isolate and to recognize each part in this mass of the materials of demolition. That can be done by the aid of the processes of direct analysis. By mentally combining these different fragments, the original building is reconstructed. This method of demolition is certainly too rough and violent. Schützenberger's operation gives us very fine fragments—small molecules of free hydrogen, of ammonia, of carbonic, acetic, and oxalic, acids which reveal extreme pulverization. These products represent about a quarter of the total mass. The other three-quarters are formed of larger fragments, the examination of which is most instructive. They belong to four groups. The first comprises five or six bodies, amido-acids or *leucins*. It proves the existence in the molecule of albumin of compounds of the series of fats—*i.e.*, arranged in an open chain. The second group is formed by tyrosin and kindred products—*i.e.*, by the bodies of the aromatic series, which force us to acknowledge the presence in the molecule of albumin of a benzene nucleus. The third group forms around the nucleus known to chemists under the name of pyrrol. The fourth comprises bodies such as the glucoproteins, connected with the sugars, or carbohydrates.

Does the fact that the molecule of albumin is destroyed in producing these compounds raise the question as to whether it implies the idea that in reality they pre-exist in it? Chemists are rather inclined to admit this. However, the conclusion does not appear to be permissible. Duclaux considers it doubtful. It is not certain that all these fragmentary bodies pre-exist in reality, and it is no more certain that a simple bringing of them together represents the primitive edifice. Materials of demolition from a house that has been pulled down give no idea of its natural architectural character. There is only one way of justifying the hypothesis, and that is to reconstitute the original molecule of albumin by bringing the fragments together. We have not got to that stage yet. The era of syntheses of such complexity is more or less near, but it has certainly not yet begun.

Moreover, it is not correct to say that the simple juxtaposition of the surfaces of fracture will reproduce the initial body. The fragments, so far as analysis has obtained them, are not absolutely what they might have been in the original structure. There they adhered the one to the other, not only by the mere contact of their surfaces of fracture, as is supposed, but in a slightly more complex manner. The fragments of the molecule are joined by bonds. We can picture them to ourselves by supposing these bonds to be like hooks. The hooks, which could only be broken by violence, are called by the chemists *satisfied atomicities.* These atomicities, set free by the breaking up, cannot remain in this condition; they must be satisfied anew. The hook tries to attach itself. In Schützenberger's experiment the addition of water provides for this necessity. A molecule of water (H_2O) splits into two, the hydrogen (H) on the one side and the hydroxyl (OH) on the other. These two elements cling to the liberated bonds of the fragments of the molecule of albumin, and thus the bodies were found complete. Schützenberger's experiment was too violent, too radical, and it gave too large a number of fragments, with their free hooks and atomicities unsatisfied, for rather a large proportion of the water added disappeared during the experiment. In one case this quantity was as much as 17 grammes per 100 grammes of albumin. The molecules of this water were employed in the reparation of the incomplete fragmentary molecules of the albumin.

It follows that Schützenberger's experiment gave too large a number of very small pieces corresponding to far too great a pulverization. The very small fragments are the molecules of acids such as acetic acid, oxalic acid, carbonic acid, molecules of ammonia, and even of hydrogen, which we know we are setting free.

But, apart from these products which represent a quarter of the molecule of albumin submitted to analysis, the other three quarters represent larger fragments which may be considered as the real constituents of the building. Thus we find four kinds of groups which may be accepted as natural. The first of these groups is that of the leucins or amido-acids. It proves the existence in the molecule of albumin of compounds of the fatty series. There is also an aromatic group—a pyridine group—and a group belonging to the category of sugars. Imagine a certain grouping of these four series. This would be the nucleus of the molecule of albumin. If we graft on to this nucleus, on to this framework as it were, so many annexes, or lateral chains, the building will be loaded with embellishments; it will have been made unstable and *ipso facto* appropriate for the part that it plays in the incessant transformations of the organism.

Kossel's Analysis. Hexonic Nucleus.—Kossel has approached the problem in another fashion. He did not attempt to attack the albumin of the egg. This body is, in fact, a heterogeneous mixture as complex as the needs of the embryo of which it forms the food. Kossel tried a physiologically simpler albuminoid. He got it from an anatomical element having no nutritive rôle, of a very elementary organization and physiological functional activity, and yet one of energetic vitality—the male generating cell. Instead of the hen's egg he therefore analyzed the milt of fish, and, in the first place, of salmon. As was to be expected from what has been said of the proteids, this living matter gives a combination of the nuclein, already known, with an albumin. The latter is abundant, forming a quarter of the total mass. Its reaction is strongly alkaline, which is the general characteristic of the variety of albumin known by the name of histones. Miescher, the learned chemist of Basle, who had noticed this basic albumin when working on the Rhine salmon, gave it the name of protamin. This is the substance submitted by Kossel to analysis in preference to the albumin of egg, so dear to the chemists who had preceded him. The disintegration of this molecule, instead of giving the series of bodies obtained by Schützenberger, gave but one, a real chemical base, *arginin*. At the first trial the albumin examined was reduced to a simple crystallizable element. The conclusion was obvious. The protamin of salmon is the simplest of albumins. To form this elementary proteid substance a hexonic base with water is all that is required.

Continuing on these lines other male generating cells were examined and a series of protamines constructed on the same type was found, and these albuminous bodies proved to be formed of a base or mixture of analogous hexonic bases: arginin, histidin, and lysin—all bodies closely akin in their properties and entirely belonging to the physical world.

Once aware of the existence of this fundamental nucleus, chemists found it in the more complex albumins where it had been missed. It was found in the albumin of egg hidden under the mass of other groups. It was recognized in all animal or vegetable albumins. The nuclei of Schützenberger may be missing. Hexonic bases are the constant and universal element of all varieties of albumins. They prevail in the chemical nucleus of the albuminous molecule, and perhaps as is suggested by Kossel, they may form it exclusively. All the other elements are superadded and accessory. The essential type of this molecular edifice, sought for so long, is known at last.

Conclusion.—To sum up, the chemical unity of living beings is expressed by saying that living matter, protoplasm, is a mixture or a complex of proteid substances with an hexonic nucleus.

CHAPTER IV.

THE TWOFOLD CONDITIONING OF VITAL PHENOMENA. IRRITABILITY.

Appearance of internal activity of the living being—Vital phenomena regarded as a reaction of the ambient world.—§ 1. Extrinsic conditions—The optimum law.—§ 2. Intrinsic conditions—The structure of organs and apparatus—How experiment attacks the phenomena of life. Generalization of the law of inertia—Irritability.

Instability. Mutability. The Appearance of Internal Activity of the Living Being.—One of the most remarkable characteristics of the living being is its instability. It is in a state of continual change. The simplest of the elementary beings, the plastid, grows and goes on growing and becoming more complex, until it reaches a stage at which it divides, and thus rejuvenated it commences the upward march which leads it once again to the same segmentation. Its evolution is thus betrayed by its growth, by the variations of form which correspond to it, and by its division.

If it be a question of beings higher in organization than the cellular element the evolutionary character of this mutability becomes more obvious. The being is formed, it grows; then in most cases, after having passed through the stages of youth and adult age, it grows old, declines and dies, and is disorganized after having gone through what we may call an ideal trajectory. This march in a fixed direction with its points of departure, its degrees, and its termination, is a repetition of the path that the ancestors of the living being have already followed.

Here, then, is a characteristic fact of vitality, or rather there are two facts. The one consists in this morphological and organic evolution, the negation of immutability, the negation of the indefinite maintenance of a permanent state or form which is regarded, on the contrary, as the condition of inert, fixed stable bodies, eternally at rest. The other consists in the repetition, realized by this evolution, of the similar evolution of its ancestors; this is a fact of heredity. Finally, evolution is always in a cycle—that is to say, that it comes to an end which brings the course of things to their point of departure.

This kind of internal activity of the living being is so striking, that not only does it serve us to differentiate the living being from the inert body, but it gives rise to the illusion of a kind of internal demon, vital force, manifested by the more or less apparent acts of the life of relation, of the motricity, of the displacement, or by the less obvious acts of vegetative life.

Vital Phenomena regarded as a Reaction of the Ambient World. Their Twofold Conditioning.—In reality, as the doctrine of energetics teaches us, the phenomena of vitality are not the effect of a purely internal activity. They are a reaction of the environment. "The idea of life," says Auguste Comte, "constantly assumes the necessary correlation of two indispensable elements:—an appropriate organism and a suitable environment. It is from the reciprocal action of these two elements that all vital phenomena inevitably result." The environment furnishes the living being with three things:—its matter, its energy, and the exciting forces of its vitality. All vital manifestation results from the conflict of two factors: the extrinsic factor which provokes its appearance; the intrinsic factor, the very organization of the living body, which determines its form. Bichat and Cuvier saw in the phenomena of life the exclusive intervention of a principle of action entirely internal, checked rather than aided by the universal forces of nature. The exact opposite is true. The protozoan finds the stimuli of its vitality in the aquatic medium which is its habitat. The really living particles of the metazoan—that is to say, its cells, its anatomical elements—meet these stimuli in the lymph, in the interstitial liquids which bathe them and which form their real external environment.

Auguste Comte thoroughly understood this truth, and has clearly expressed it in the passage we have just quoted. Claude Bernard has fully developed it and given it its classical form.

In order to manifest the phenomena of vitality, the elementary being, the protoplasmic being, requires from the external world certain favourable conditions; these it finds there, and they may be called the stimuli, or extrinsic conditions of vitality. This being possesses no initiative or spontaneity in itself, it has only a faculty of entering into action when an external stimulus provokes it. This subjection of the living matter is called *irritability*. The term expresses that life is not solely an internal attribute, but an internal principle of action.

§ 1. Extrinsic Conditions.

Extrinsic Conditions.—By showing that every vital manifestation results from the conflict of two factors: the extrinsic or physico-chemical conditions which determine its appearance, and the intrinsic or organic conditions which regulate its form, Claude Bernard dealt a mortal blow at the old vitalist theories. For he has not only asserted the close dependence of the two kinds of factors, but he has shown them in action in most physiological phenomena. The study of the extrinsic or physico-chemical conditions necessary to vital manifestations teaches us our first truth—namely, that they are not infinitely varied as might be supposed. They present, on the contrary, a remarkable uniformity in their essential qualities. The fundamental conditions are the same for the animal or vegetable cells of every species. They are four in number:—*moisture*, the air, or rather *oxygen*, *heat*, and a certain *chemical constitution* of the medium, and the last condition, the enunciation of which seems vague, becomes more precise if we look at it a little closer. The chemical constitution of media favourable to life, the media of culture, obeys three general laws. It is the knowledge of these laws which formerly enabled Pasteur, Raulin, Cohn, and Balbiani to provide the media appropriate to the existence of certain relatively simple organisms, and thus to create an infinitely valuable method for the study of nutrition, etc.,—namely, the *method of artificial cultures*, numerous developments of which have been shown us by microbiology and physiology.

The Optimum Law.—It has been said, and it is more than a play on words, that the conditions of the vital medium were the conditions of the *juste milieu*. Water is wanted, there must not be too much or too little. Oxygen is necessary, and also in certain proportions. Heat is required, and for that, too, there is an optimum degree. Certain chemical compounds are needed and, in this respect too, there must also be *optima* proportions.

Water is a constituent element of the organisms. They contain fixed proportions for the same tissue, proportions varying from one tissue to another (between $\frac{2}{3}$ and $\frac{9}{10}$). The cell of a living tissue requires around it an aqueous atmosphere, formed by the different juices of the organism, the interstitial liquids, the blood, and the lymph. We are deceived by appearances when we distinguish between aerial, aquatic, and land-dwelling animals, and when we speak of the air, the water, and the land as their natural environment. If we go to the bottom of things, and fix our attention on the real living unities, on the cells of which the organism is composed, we shall find around them the juices, rich in water, which are their real environment. If these juices are diluted or concentrated the least in the world, life stops. The cell, the whole animal, falls into a state of latent life, or dies. "All living beings are aquatic," said Claude Bernard. "Beings that live in the air are in reality wandering aquariums," said another physiologist. "No moisture, no life," wrote Preyer. The environment must contain water, but it must contain it in certain proportions. In the higher animals there is a mechanism which works automatically to keep at a constant level the quantity of water in the blood. Researches on the lavage of the blood (A. Dastre and Loye) have clearly shown this.

Oxygen is also necessary to life. It is the *pabulum vitæ.* But the discovery of the beings called by Pasteur *anaerobia* appears to contradict this statement. Pfeffer, the illustrious botanist, was certain, in 1897, that the dogma of the necessity of oxygen no longer held good. This is no longer tenable. In 1898 Beijerinck carried out most careful researches on anaerobia said to have been cultivated in a vacuum, such as the *bacteria of tetanus* and the *septic vibrion;* or on those to which oxygen seems to be a poison, such as the *butyric* and the *butylic ferments*, the anaerobia of putrefaction, the reducing spirilla of the sulphates. All use free oxygen. They consume very little it is true; they are micro-aerobia. The other organisms, on the contrary, need more. They are macro-aerobia or simply *aerobia.* Besides, if the so-called anaerobia take little or no free oxygen, it matters little. They take the oxygen in combination. It may be said with L. Errera that they have an affinity for oxygen, for they extract it from its combinations, and that "they are so well adapted to this mode of existence that life in the open air being too easy no longer suits them." There are for the different animal species different optima of oxygen.

Living beings require a certain amount of heat. Life, which could not have existed on the globe when it was incandescent, will not be able to exist when it is frozen. For each organism and each function there is a maximum and a minimum of temperature compatible with activity. There is also an optimum. For instance, the optimum is 29° C for the germination of corn.

The condition of the optimum exists in the same

way for the chemical composition of the vital medium —and for the other ambient physical conditions, such as atmospheric pressure.

It is therefore a law of *universal* scope, a regulating law, as it were, of life. Life is a function of extrinsic variables, water, air, heat, the chemical composition of the medium, and pressure. "Every vital phenomenon begins to be produced, starting from a certain stage of the variable (minimum), becomes more and more vigorous as it increases up to a determinate value (optimum), weakens if the variable continues to increase, and disappears when it has reached a certain limiting value (maximum)." This law, proved by Sachs, the German botanist, in 1860, apropos of the action of temperature on the germination of plants, by Paul Bert in 1875, apropos of the action of oxygen and of atmospheric pressure on animals, and already formulated at that time by Claude Bernard, was illustrated by Leo Errera in 1895. It is a law of moderation. It expresses La Fontaine's "*rien de trop*," Terence's "*ne quid nimis*," the μηδὲν ἄγαν of Theognis, and the biblical phrase "*omnia in mensura et numero et pondere.*" L. Errera sees the profound cause of this optimum law in the properties of the living protoplasm, which are mean properties. It is semi-liquid. It is composed of albuminoid substances, which can stand no extremes either from the physical or from the chemical points of view.

§ 2. INTRINSIC CONDITIONS. THE LAW OF THE CONSTITUTION OF ORGANS AND APPARATUS.

Law of the Constitution of Organs and Apparatus.— If we consider more highly organized beings, the influence of the intrinsic conditions appears quite as clearly. As we have seen, this is so that the requisite fundamental materials may be spent by each element in suitable proportions,—water, chemical compounds, air, and heat,—that organs may be added to organs, and that apparatus may be set to work in complex structures. Why a digestive apparatus? To prepare and introduce into the internal medium liquid materials which are necessary to life. Why a respiratory apparatus? To impart the vital gas necessary to the cells, and to expel the gaseous excrement, the carbonic acid which they reject. Why a circulatory apparatus? To transport and renew this medium throughout. The apparatus, the functional wheels, the vessels, the digestive and respiratory mechanisms do not exist for themselves, like the random sketches of an artistic nature. They exist for the innumerable anatomical elements which people the economy. They are arranged to assist and more rigorously to regulate cellular life with respect to the extrinsic conditions which it demands. They are, in the living body, as in civilized society, the manufactories and the workshops which provide for the different members of society dress, warmth. and food. In a word, the *law of the construction of organisms* or of the *bringing to perfection of an organism* is the same as the law of cellular life. It is otherwise suggestive as the law of *division of physiological labour* formerly enunciated by Henry Milne-Edwards; and in every case it has a more concrete significance. Finally, it brings the organic functional activity into relation with the conditions of the ambient medium.

How Experiment acts on the Phenomena of Life.—

The two orders of conditions, the one provided by the being itself, the other by external agents, are equally indispensable—and therefore of equal importance or dignity. But they are not equally accessible to the experimentalist. It is not easy to exercise on the organization direct and measurable actions. On the contrary, the physical conditions are in the hands and at the discretion of the experimenter. By them he may reach the vital manifestations as they appear, stimulate or check them, defer or precipitate them. Thus, for instance, the physiologist suspends or re-establishes at his will full vital activity in a multitude of reviviscent or hibernating beings, such as grains, the infusoria capable of *encystment*, the vibrio, the tardigrade, the cold-blooded animals, and perennial plants.

The ambient world therefore furnishes to the animal and to the vegetable, whole or fragmentary, those materials of its organization which are at the same time the stimuli of its vitality. That is to say, the vital mechanism would be a dormant and inert mechanism if nothing in the surrounding medium could provoke it to action or give it a check. It would be a kind of steam engine without coal and fire.

Living matter, in other words, does not possess real spontaneity. As I have shown elsewhere, the law of inertia which it is supposed it obeys with inert bodies is not special to them. It is applied to the living bodies whose apparent spontaneity is only an illusion contradicted by physiology as a whole. All the vital manifestations are responses to a stimulus of acts provoked, and not of spontaneous acts.

Generalization of the Law of Inertia in Living Bodies. Irritability.—In fact, vulgar prejudice opposes this view. The opinion of the average man distrusts it. It applies the law of inertia only to inert matter. This is because the vital response does not always immediately succeed the external stimulus, and is not always proportional to it. But it is sufficient to have seen the flywheel of a steam engine to understand that the restitution of a mechanical force cannot be instantaneous. It is sufficient to have had a finger on the trigger of a firearm to know that there is no necessary proportion between the intensity of the stimulus and the magnitude of the force produced. Things happen in the living just as in the inert machine.

The faculty of entering into action when provoked by an external stimulus has received, as we have said, the name of *irritability*. The word is not used of inert matter. However, the condition of the latter is the same. But there is no need to affirm its irritability, because no one denies it. We know perfectly well that brute matter is inert, that all the manifestations of activity of which it is the theatre are provoked. Inertia is for it the equivalent of irritability in living matter. But while it is not necessary to introduce this idea into the physical sciences, where it has reigned since the days of Galileo, it was, on the contrary, necessary to affirm it in biology, precisely because it was in biology that the opposing doctrine of vital spontaneity ruled supreme.

Such was the view held by Claude Bernard. He never varied on this point. *Irritability*, said he, is the property possessed "by every anatomical element (that is to say, the protoplasm which enters into its constitution) of being stimulated into activity and of reacting in a certain manner under the influence of the external stimuli." He could not claim that this was a distinguishing characteristic between living bodies and brute bodies, and that all the less because he always tried to efface on this point the distinctions which were current in his time, and which were established by Bichat and Cuvier. And so also Le Dantec does not seem to have thoroughly grasped the ideas of the celebrated physiologist on this point when he asserts, as if he were thereby contradicting the opinion of Claude Bernard and his school, that irritability is not something peculiar to living bodies.[1]

[1] These ideas are clearly brought to light in a series of articles in the *Revue Philosophique*, published in 1879 under the title of "La problème physiologique de la vie," and endorsed by A. Dastre in his commentary on the *Phénomènes communs aux animaux et aux plantes.*

CHAPTER V.

THE SPECIFIC FORM. ITS ACQUISITION. ITS REPARATION.

§ 1. Specific form not special to living beings—Connected with the whole of the material conditions of the body and the medium—Is it a property of chemical substance?—§ 2. Acquisition and re-establishment of the specific form—Normal regeneration—Accidental regeneration in the protozoa and the plastids—In the metazoa.

§ 1. The Specific Form.

The Specific Form is not Peculiar to Living Beings.—The position of a *specific form*—the acquisition of this typical form progressively realized—the re-establishment when some accident has altered it—these are the features which we consider distinctive of living beings, from the protophytes and the lowest protozoa to the highest animals. Nothing gives a better idea of the unity and the individuality of the living being than the existence of this typical form. We do not mean, however, that this characteristic belongs to the living being alone, and is by itself capable of defining it. We repeat that this is not a case with any characteristic. In particular the *typical form* belongs to crystal as well as to living beings.

The Specific Form depends on the sum of Material Conditions of the Body and the Medium.—The consideration of mineral bodies shows us form dependent on the physico-chemical conditions of the body and the medium. The form depends mainly on physical conditions in the cases of a drop of water falling from a tap, of the liquid meniscus in a narrow tube, of a small navel-shaped mass of mercury on a marble slab, of a drop of oil "emulsioned" in a solution, and of the metal which is hardened by hammering or annealed. In the case of crystals the form depends more on chemical conditions. It is crystallization which has introduced into physics the idea that has now become a kind of postulate—namely, that the specific form is connected with the chemical composition. However, it is sufficient to instance the dimorphism of a simple body, such as sulphur, sometimes prismatic, sometimes octahedric, to realize that substance is only one of the factors of form, and that the physical conditions of the body and of the medium are other factors quite as influential.

Is the Specific Form a Property of the Chemical Substance?—How much truer this restriction would be if we consider, instead of a given chemical compound, an astonishingly complex mixture, such as protoplasm or living matter, or the more complex organism still—the cell, the plastid.

Are there not great differences between the substance of the cellular protoplasm, or cytoplasmic substance, and that of the nucleus? Should we not distinguish in the former the hyaloplasmic substance; the microsomic in the microsomes; the linin between its granulations; the centrosomic in the centrosome; the archoplasmic in the attraction sphere; not to mention the different leucins, the vacuolar juice, and the various inclusions? And in the nucleus must we not consider the nuclear juice, the substance of the chromosomes, and that of the nucleoles? And is not each of these probably a very complex mixture?

However, it is to this mixture that we attribute the possession of a form, in virtue of and by extension of the principles of crystallization, which definitely teach us that these mixtures cannot have form; that form is the attribute of pure bodies, and is only obtained by the separation of the blended parts—*i.e.*, by a return to homogeneity. There are therefore very good reasons for hesitating before we transfer the absolute principle of the dependence between chemical form and composition, as some philosophical biologists have done, from the physical sciences—where it is already subject to serious restrictions—to the biological sciences.

Le Dantec, however, has made this principle the basis of his biological system. He therefore finds in the crystal the model of the living being. He thus gives a physical basis to life.

Is it a question in this system of explaining this incomprehensible, this unfathomable mystery, which shows the egg cell attracting to itself materials from without and progressively building up that amazing structure which is the body of the animal, the body of a man, of any given man, of Primus, for example? It is said that the substance of Primus is specific. His living substance is his own, special to him; and that, too, from the beginning of the egg to the end of its metamorphosis. It only remains to apply to this substance the postulate, borrowed from crystallography, of the absolute dependence of the nature of substance on the form it assumes. The form of the body of the animal, of the man, of Primus, is the crystalline form of their living substance. It is the only form of equilibrium that this substance can assume under the given conditions, just as the cube is the crystallized form of sea salt, the only state of equilibrium of chloride of sodium in slowly evaporated sea water. Thus the problem of the living form is reduced to the problem of the living substance, which seems easier; and at the same time the biological mystery is reduced to a physical mystery. It is clear that this way of looking at things simplifies prodigiously—and, we must add, simplifies far too much—the obscure problem of the relation of form to substance, simultaneously in the two orders of science. This may be summed up in a single sentence: There is an established relation between the specific form and the chemical composition: the chemical composition *directs* and implies the specific form.

We need not now examine the basis of this opinion. If it is nothing but a verbal simplification, a unification of the language applied to the two orders of phenomena, it implies an assimilation of the mechanisms which realize them. To the organogenic forces which direct the building up of the living organisms it brings into correspondence the crystallogenic forces which group, adjust, equilibrate, and harmonize the materials of the crystal.

When it is a question of the application of a principle such as this, in order to test its legitimacy we must always return to the experimental foundations. Let us imagine, for example, a simple body, such as sulphur, heated and brought to a state of fusion—that is to say, homogeneous, isotropic, in an undisturbed medium the only change in which will be a very gradual cooling down. These are the typical crystallogenic conditions. The body would take a given crystalline form. It is from experiments such as this that we derive the idea of *a specific form connected with a chemical constitution.*

But in drawing this conclusion our logic is at fault. The real interpretation suitable to this case, as in all others, is that the specific form is suitable to the substance, and also to the physical, chemical, and mechanical conditions in which it is placed. And the proof is that this same substance, sulphur, which takes the prismatic form immediately after fusion, will not retain that form, but will pass on to the quite different octahedral form.

It is so with the specific form of the living being—that is to say, with the assemblage of its constituent materials co-ordinated in a given system—in a word, with its organization. This is suitable to its substance, and to all the material, physical, chemical, and mechanical conditions in which it is placed. This form is the condition of material equilibrium corresponding to a very complex situation, to a sum of given conditions. The chemical condition is only one of these. And further, it is hardly proper to speak of a "chemical substance" when we refer to an astonishingly complex mixture which is in addition variable from one point to the other of the living body. When we thus reduce phenomena to their original signification, false analogies disappear. To say with Le Dantec that the form of the greyhound is the condition of equilibrium of the "greyhound chemical substance" is saying much; and too much, if it means that the body of the greyhound has a substance which behaves in the same way as homogeneous, isotropic masses like melted sulphur and dissolved salt. It were better to say much less, if it means, as it will in the minds of the physiologists, that the body of the greyhound is the condition of equilibrium of a heterogeneous, anisotropic, material system, subjected to an infinite number of physical and chemical conditions.

The idea of connecting form, and by that we mean organization, with chemical composition did not arise in the minds of chemists or physiologists. Both have expressed themselves very clearly on this point.

"We must distinguish," said Berthelot, "between the formation of the chemical substances, the assemblage of which constitutes organized beings, and the formation of the organs themselves. This last problem does not come into the domain of chemistry. No chemist will ever claim to have formed in his laboratory a leaf, a fruit, a muscle, or an organ. . . . But chemistry has a right to claim that it forms direct principles—that is to say, the chemical materials which constitute the organs." And Claude Bernard in the same way writes:—"In a word, the chemist in his laboratory, and the living organism in its apparatus, work in the same way, but each with its own tools. The chemist can make the products of the living being, but he will never make the tools, because they are the result of organic morphology."

§ 2. The Acquisition and Re-establishment of the Specific Form.

Acquisition of the Typical Form.—The acquisition of the typical form in the living being is the result of ontogenic work which cannot be examined here. In the elementary being, the plastid, this work is blended with the work of nutrition. It is *directed nutrition.* It consists of a simple increase from the moment the element is born by the division of an anterior element, and of a necessarily restricted differentiation. It is a rudimentary embryogeny. In the complex being, metazoan or metaphyte, the organism is constituted, starting from the egg, by the growth, by the bipartition of the elements, and their differentiation, accomplished in a certain direction and in conformity with a given plan. This, again, is directed nutrition, but here the embryogeny is complex. The directing plan of operations is no doubt the consequence of the material conditions realized each moment in the organism.

Normal Regeneration.—Not only do living beings themselves construct their typical architecture, but they re-establish it and continually reconstitute it, according as accidents, or even ordinary circumstances, tend to destroy it; in a word, they become rejuvenescent. This regeneration consists in the reformation of the parts that are altered or carried away in the normal play of life, or by the accidents which disturb its course.

Thus there is a *normal physiological regeneration,* which is, so to speak, the prolongation of the ontogenesis—*i.e.*, of the work of formation of the individual. We have examples in the reconstitution of the skin of mammals—in the throwing off of the epidermic products constantly used up in their superficial and distal parts and regenerated in their deeply-seated parts; in the loss and the renewal of teeth at the first dentition and in certain fish in the fact of successive dentitions; in the periodical renewal of the integument in the larvæ of insects, and in the crustaceæ; and finally in the destruction and the neo-formation of the globules of the blood of vertebrates, of the glandular cells, and of the epithelial cells of the intestine.

Accidental Regeneration in Protozoa and Plastids.—There is also an *accidental regeneration* which more or less perfectly renews the parts that are lost. This regeneration has its degrees, from the simple cicatrization of a wound to the complete reproduction of the part cut off. It is very unequally developed in zoological groups even when they are connected. In the elementary monocellular beings—*i.e.*, in the anatomical elements and in the protozoa,—the experiments in merotomy, *i.e.*, in *partial section*, enable us to appreciate the extent of this faculty of regeneration. These experiments, inaugurated by the researches of Augustus Waller in 1851, were repeated by Gruber in 1885, continued by Nussbaum in 1886, Balbiani in 1889, Verworn in 1891, and have been reproduced by a large number of observers. They have shown that the two fragments cicatrize, and are repaired, building up an organism externally similar to the primitive organism, but smaller. The two new organic units do not, however, behave in the same way. That which retains the nucleus possesses the faculty of regeneration, and of living as the primitive being lived. The protoplasmic fragment, which does not contain the nucleus, cannot rebuild this absent organ; and though it has functional activity in most respects, just as the nucleated fragment, yet it is distinguished from it in others of great importance. The anucleated fragment of an infusorian behaves as the nucleated, and as the whole animal so far as the movements of the body, the cilia, prehension of food, evacuation of fæces, and the rhythmical contraction of the pulsatile vesicules are concerned. But Balbiani's researches in 1892 have shown us that secretion, complete regeneration, and the faculty of reproduction by fission can take place only in the nucleated fragment—*i.e.*, in the nucleus.

Accidental Reproduction in the Metazoa.—Among multicellular beings the faculty of reproduction is met with in the highest degree in plants, where we find it in the process of propagation by slips. In animals it is the most marked in Cœlenterata. Trembley's experiments are a striking instance. We know that when the hydra is cut into tiny pieces it reproduces exactly as many complete beings. Among the worms, Planaria afford a similar example. Every fragment, provided its volume is not less than a tenth of that of the whole, can reproduce a complete, entire being. The snail can produce a large part of its head, including the tentacles and the mouth. Among the Tritons and the Salamanders the faculty of regeneration reproduces the limbs, the tail, and the eye. In the Frog family, on the contrary, the work of regeneration does not go beyond cicatrization, and it is the same with Birds and Insects.

It is really startling to see in a vertebrate like the Triton the stump of an arm with its fragment of humerus reproducing the forearm and the hand in all their complexity, with their skeleton, blood vessels, nerves, and teguments. We say that the limb has *budded*, as if there were a germ of it which develops like the seed of a plant, or as if each transverse portion of the limb, each slice, so to speak, could re-form the slice that follows.

The mechanism of generation and that of regeneration alike raise problems of the highest importance. Does the part become regenerated just as it was formed at first? Does the regeneration repeat the ontogeny? Is it true that a lost organ is never regenerated (the kidney for instance)? Does the symmetrical organ enjoy a compensating and hypertrophic development, as Ribbert has asserted? And further, if the organ be removed and transplanted to another position, can it be grafted there, as Y. Delage maintains? These are very important questions; but if we dwell upon them, we shall be diverted from our immediate object. Our task is to look at these facts from the point of view of their significant and characteristic meaning in vitality. Flourens invoked on their behalf the intervention of vital forces, *plastic* and *morphoplastic*. But, as we shall see later, these phenomena of cicatrization, of reparation, of regeneration, these more or less complete efforts for the re-establishment of the specific form, although they are found in all living beings in different degrees, are not exclusively confined to them. We find them again in some representatives of the mineral world—in crystals, for instance.

CHAPTER VI.

NUTRITION.

FUNCTIONAL ASSIMILATION. FUNCTIONAL DESTRUCTION. ORGANIC DESTRUCTION. ASSIMILATING SYNTHESIS.

The extreme importance of nutrition—§ 1. Effect of vital activity—Destruction or growth—Distinction between the living substance and the reserve-stuff mingled with it—Organic destruction—Destruction of reserve-stuff—Destruction of living matter—Growth of living matter—§ 2. The two categories of vital phenomena—Foundations of the idea of functional destruction—The two kinds of phenomena of vitality—Criticism of Claude Bernard—Current views—Criticism of Le Dantec's new theory of life.—§ 3. Correlation of the two kinds of vital facts—Law of connection—Contradictions in the new theory.—§ 4. The characteristics of nutrition—Its definition—Its permanence—Erroneous idea of the vital vortex—Formative assimilation of reserve-stuff—Formative assimilation of protoplasm—Death, real and apparent.

The Immense Importance of Nutrition.—We now come to the important feature of vitality. All other characteristics of living matter, its unstable equilibrium, its chemical and anatomical organization, the acquisition and the maintenance of a typical form, are only secondary properties, so to speak, subordinate with reference to *nutrition.* Generation itself is only a mode. *Nutrition* is the essential attribute of life. It is life itself.

Before we define it a few preliminary explanations are necessary.

The most striking thing in living matter is its *growth.* An animal, a vegetable, is something which is first more or less minute, and which grows. Its characteristic is to expand—from the spore, the seed, the slip, the egg—it grows.

Whether we are dealing with a cellular element, a plastid, or a complex being, their condition is the same in this respect. No doubt when the animal or plant has reached a certain stage of development its growth is stopped, and for a more or less lengthy period it remains in the adult stage, in what seems to be equilibrium. But even then there is no check in the manufacture of living matter; there is only a compensation between its production and its destruction.

It is important to reduce to order the ideas on this important subject, which at present are confused, inconsistent, and contradictory. In biology grievous confusion reigns.

§ 1. Effect of the Vital Activity. Destruction or Growth?

Distinction between the Living Substance and the Reserve-stuff mingled with it.—The physiology of nutrition has given rise to a vast body of research during the last half-century. Physiological schools, masters and pupils, such as the school at Munich under Voit and Pettenkofer, Pflüger's at Bonn, Rubner's, and those of Zuntz and von Noorden at Berlin, and a large number of zootechnical and agricultural laboratories through the whole world have for years past been engaged in analyzing ingesta and egesta, in drawing up schedules of nutrition, in order to determine the course of decomposition and reconstitution of the living material.

If I were asked what, in my opinion, is the most general result of all this labour, I would reply that it has affirmed and corroborated the important distinction which must be drawn between *living substance, properly so called*, and *reserve-stuff*. The latter, the *reserve-stuff* of albuminoids, carbohydrates, and fats, are so intimately intermingled with the living substance that they are in most cases very difficult to distinguish from it.

Organic Destruction.—A second point, which is placed equally beyond doubt, is that the vital functional activity is accompanied by a destruction of the immediate principles of the organism, in the direction of their simplification. This functional destruction cannot be doubted in the case of differentiated organs in which the functional activity is evident, intermittent, and in some measure distinct from the other vital phenomena which take place in them. For example, in the case of contracting muscles the respiratory carbonic acid and urinary carbon are the irrefutable proofs of this destruction: weak in repose, abundant during activity, and in proportion to it. There can be no doubt on this point. The truth laid down by Claude Bernard under the name of the *law of functional destruction* has been doubly consecrated by experiment and theory. According to the energetic theory, in fact, mechanical and thermal energies manifested in the vital functional activity can only have their source in the chemical energy set free by the destruction of the immediate principles of the organism, reduced to a lower degree of complexity.

Destruction of Reserve-stuff.—But now the disagreement begins. What are these decomposed, destroyed principles? Do they belong to the cellular reserve-stuff or to the living matter properly so called? There is no doubt that most of them belong to the reserve-stuff. For example, this is especially true of glycogen, which is consumed in muscular contraction just as coal is consumed in the furnace of the locomotive; and glycogen is a reserve-stuff of muscle. These reserve-stuffs destroyed in the functional activity can be built up again only during repose.

But it is not yet certain whether the living matter itself, the active protoplasm, the muscular protoplasm, takes part in this destruction, whether it provides it with elements. Experiments have proved contradictory. Experimenters have isolated the nitrogenous wastes (urea) after muscular labour, and they have compared them with the wastes of the period of repose. These nitrogenous wastes bear witness to the destruction of albuminoid substances, and the latter are the constituent principles of living matter. If—under conditions of sufficient alimentation—the muscular functional activity involves more nitrogenous waste, *i.e.*, a greater destruction of albuminoids, it might be supposed that the living material properly so called has been used up and destroyed for its own purposes. (And here again there might be a reserve-stuff of albuminoids, distinct from the living protoplasm itself, and more or less incorporated with it.)

But experiment so far has not given decisive results. The latest experimental researches, such as those of

Igo Kaup, of Vienna, which date from 1902, tell us as uncertain a tale as their predecessors. The increase in the destruction of albumen has not been constant; the conditions of the observations do not justify our making an assertion either *pro* or *con*.

Destruction of Living Matter.—As no certain answer is supplied by experiment, theory intervenes and gives two conflicting answers. The majority of physiologists are inclined to believe in *the destruction of the living substance as the result of its own functional activity.* The functional activity would therefore destroy not only the reserve-stuff, but also the protoplasmic material. This is the current view. Only this opinion is strongly challenged by the positive teaching of science. It is certain that this material, in the muscle, is but little attacked, if it is attacked at all. We have seen above that the physiologists, with Pflüger and Chauveau, are agreed on this point. The vital functional activity in particular is destructive to the reserve-stuffs. It does not destroy them much; it destroys the organic material still less. Both would be repaired in functional repose.

Growth of Living Matter.—The second assertion is diametrically opposed to this. Not only, says Le Dantec, is the muscle not destroyed in the functional activity, but it grows. Contrary to universal opinion, the protoplasmic material increases by activity, and it is destroyed in repose. There would thus be a general law—*the law of functional assimilation.* "A cell of brewers' yeast when introduced into a sugared must makes this must ferment, and at the same time, so far from destroying it, it increases it. Now, the fermentation of the must is exactly the same as the functional activity of the yeast." It is, says the same author, a mistake to believe that the phenomena of functional activity, of *vital activity*, only takes place at the price of organic destruction. Here, then, are these two competing views. They are not so very far apart as a matter of fact, since the question at issue is one of deciding between a slight destruction and a slight growth, but theoretically they are strongly opposed. Moreover, they are arbitrary, and *experiment* has not decided between them.

§ 2. THE TWO CATEGORIES OF VITAL PHENOMENA.

Foundation of the Idea of Functional Destruction. Claude Bernard.—The doctrine of functional destruction has been laid down with remarkable power by Claude Bernard. But the terms in which he has expressed it in a measure betray the thoughts of the great physiologist, or, at any rate, overstep the immediate fact he had in view. "The phenomena of destruction are very obvious. When movement is produced, when the muscle contracts, when volition and sensibility are manifested, when thought is exercised, when the gland secretes, then the substance of the muscles, of the nerves, of the brain, of the glandular tissue, becomes disorganized, destroyed, and consumed. So that every manifestation of a phenomenon in the living being is necessarily connected with an organic destruction." To Claude Bernard organic destruction is a truth. To Le Dantec it is an error. Which is right? Clearly Claude Bernard. He bases his conviction on the analyses of the materials excreted in the process of physiological work. The excreta bear witness to a certain organic demolition. Generalizing this teaching of experiment the illustrious biologist divined the fundamental law of energetics before the idea of energetics had made much way in France. Every act which expends energy, which produces heat or motion, any manifestation whatever that may be looked upon as an energetic transformation, necessarily expends energy, and that energy is borrowed from the substance of the organism. These substances are simplified, broken up, and destroyed. Now the functional activity of the muscle produces heat and movement in warm-blooded as well as in cold-blooded animals. The functional activity of the glands produces heat, as has been shown by the celebrated experiments of C. Ludwig on the salivary secretion, and as is also shown by the study of thermal topography in the vertebrates. The functional activity of the nerves and the brain produces a slight quantity of electricity and heat, as most observers have agreed. The functional activity of the electrical and of the photic apparatus also expends energy. Finally, the eye which receives the photic impression destroys the purple matter of the retina, and that purple matter, as we well know, is recuperated in the dark during the repose of the organ. Everything that is expressed objectively, everything that is a phenomenon in the living being—with the exception of growth and formation, which are generally slow phenomena, and of which we can only get an idea by the comparison of successive states—all these energetic manifestations suppose a destruction of organic matter, a chemical simplification, the source of the energy manifested. And that is why material destruction does not merely coincide with functional activity, but is its measure and expression.

The Two Kinds of Phenomena of Vitality.—Another point on which Claude Bernard is right and his opponent is wrong is not less fundamental. What are we to understand by functional phenomena? This is the very point at issue. Now, in the mind of physiologists, this expression has a perfectly definite meaning. It is not so with Le Dantec. Physiologists who have studied animals rather high in organization —in which the differentiation of phenomena enables us to grasp the fundamental distinction—have readily recognized that the phenomena of living beings are divided into two categories. There are some which are intermittent, alternative, which take place, or grow stronger at certain moments, but which cannot be continuous—they are the *functional acts;* there are others in which this characteristic of explosives, energetic expenditure and intermittence, do not appear—they are, in general, the *nutritive acts.* The muscle which contracts shows functional activity. It has an activity and a repose. During this apparent repose we must not say that it is dead; it has a life, but that life is obscure as far as the salient fact of functional movement is concerned. The salivary gland which throws up waves of saliva when the food is introduced and masticated in the mouth, or when the chord of the tympanum is at work, is in a state of functional activity; this is the salient phenomenon. But before, though nothing, absolutely nothing, was flowing through the glandular canal, yet the gland was not reduced to the condition of a dead organ: it was living a more obscure, a less evident life. The microscopical researches of Kühne, Lea, and Langley, now universally verified, show us that during this time of apparent repose the cells were loading up their granulations and getting ready the materials of secretion, as just now the muscle at rest was accumulating glycogen and the reserve-stuff which are to be expended and destroyed in contraction. Similarly, with regard to the functional activity of the other glands, of the brain, etc. Claude Bernard was, therefore, perfectly right, when he took as his model the chemists who distinguished between exothermic and endothermic reactions, and who classed the phenomena of life into two great divisions: those of functional activity, and those of functional repose.

1st. *The phenomena of functional activity* "are those which 'leap to the eyes,' and by which we are inclined to characterize life. They are conditioned by the effects of wear and tear, of chemical simplication, and of the organic destruction which liberates energy." And it must be so, because these functional manifestations expend energy. These phenomena, which are the most obvious, are also the least specific phenomena of vitality. They form part of the general phenomenality.

2nd. The *phenomena* which accompany *functional repose* correspond to the building up of the reserve-stuff destroyed in the preceding period, to the organizing synthesis. The latter remains "internal, silent, concealed in its phenomenal expression, noiselessly gathering together the materials which will be expended. We do not see these phenomena of organization directly. The histologist and the embryogenist alone, following the development of the element or of the living being, sees the changes and the phases which reveal this silent effort. Here is a store of substance; there, the formation of an envelope or a nucleus; there, a division or multiplication, a renewal." This type of phenomena is the only type which has no direct analogues: it is peculiar, special to the living being: what is really vital is this evolutive synthesis. Life is creation.

Criticism of Claude Bernard.—All this is perfectly true. Thirty years of the most intensive scientific development have run by since these lines were written, and have not essentially changed the ideas therein expressed. His work in its broad lines remains intact. Does that imply, however, that everything is perfect in detail and expression, and that there is no reason for making it more precise or for giving it fresh form? No doubt this is not so. Although Claude Bernard contributed to establish the essential distinction between the real living protoplasm and the materials of reserve-stuff which it contains, he has not drawn a sufficiently clear distinction between what belongs to each of the categories. He has not specified, in relation to organic destruction, what bearing it has on the organic materials of reserve-stuff. Sometimes he uses the term "organic destruction," which is correct, and sometimes "vital destruction," which is of doubtful import. Further, he employs an obscure and paradoxical formula to characterize the obvious but nevertheless not specific phenomena of organic destruction, and he says: "life is death."

Current Views.—Nowadays, if I may express a personal opinion on this important distinction between functional activity and functional repose, I should say that after having distinguished between the two categories of phenomena we must try to correlate them. We must try to discover, for instance, what there is in common between the muscle in repose and the muscle in contraction, and to perceive in the *muscular tonus* a kind of bridge thrown between these two conditions. The functional activity would be uninterrupted, but it would have its degrees of activity. The muscular tonus would be the permanent condition of an activity which is capable only of being considerably raised or lowered. Similarly for the glandular functional activity; the periods of charge must be connected with the periods of discharge. In a word, following the constant path of the human mind in scientific knowledge, after having drawn the distinctions that are necessary to our understanding of things, we must obliterate them. After having dug our ditches we must fill them up again. After having analyzed we must synthesize. The distinction between the phenomena of *functional activity* and the phenomena of *functional repose* or *purely vegetative* and nutritive *activity*, though only valid in the case of a provisional and approximate truth, none the less throws light on the obscure regions of biology.

The succession of energy and repose, of sleep and awakening, is a universal law, or at least a very general law, connected with the laws of energetics. The heart, the lungs, the muscles, the glands, the brain obey in the most obvious manner this obligation of rhythmical activity. The reason is clear. It is because the functional activity involves what is generally a sudden expenditure of energy, and this has to be replaced by what is generally a slow process of reparation. Functional activity is an explosive destruction of a chemical reserve which is built up again more or less slowly.

Criticism of Le Dantec's "New Theory of Life."—Let us now examine the antithesis of Claude Bernard's views. There are evidently rudimentary organisms in which the differentiation of the two categories of phenomena is but little marked; in which, apart from the movement, it is impossible to recognize intermittent, functional activities clearly distinct from morphogenic activity. It is not in this domain of the indistinct that we must seek the touchstone of physiological distinctions. Clearly, we must not choose these elementary plastids to test the doctrine of functional assimilation and functional destruction. But is not this exactly what Le Dantec did when he began his researches on brewers' yeast? When we try to examine things, we must choose the conditions under which they are differentiated, and not those in which they are confused. And this is why, in the significant words of Auguste Comte, "the more complex living beings are, the better known they are to us." The philosopher goes still farther in this direction, and adds "directly it is a question of the characteristics of animality we must start from the man, and see how those characteristics are little by little degraded, rather than start from the sponge and endeavour to discover how these characteristics are developed. The animal life of the man assists us to understand the life of the sponge, but the converse is not true."

When, moreover, we consider a vegetable organism such as yeast, which derives its energy, not from itself, not from the potential chemical energy of its reserve-stuff, but directly from the medium—that is to say, from the potential chemical energy of the compounds which form its medium of culture,—we then find ourselves in the worst possible situation for the recognition of organic destruction. Further, it is doubly wrong to assert that in so ill-chosen a type the functional phenomena do not result from an organic destruction—for at first there are no very distinct functional phenomena here—and, in the second place, there certainly is organic destruction. The phenomena of the morphogenic vitality detected in the yeast are the exact concomitants, or the results, of the destruction of an organic compound, which in this case is sugar. The yeast destroys an immediate principle, and this is the point of departure of its vital manifestations; only, it has not, as a preliminary, clearly incorporated and assimilated this principle. When, therefore, the functional phenomena are effaced and disappear, we none the less find phenomena of destruction of organic compounds which are in a measure, a preface to the phenomena of growth. This is what happens in the case of brewers' yeast: and here, again, the two categories of facts exist. Once more, we find, in the first place, the phenomena of destruction (destruction of sugar, reduced by simplification to alcohol and carbonic acid)—phenomena which this time no longer respond to obvious functional manifestations; and, in the second place, the phenomena of chemical and organogenic synthesis, corresponding to the growth of the yeast and the multiplication of its protoplasm. The former are no longer detected, as we have just said, by striking manifestations. However, it is not true that everything which is visible and which may be isolated outside the activity of the yeast is part of those phenomena. The boiling of the juice or the mash, the heat given off by the copper, all this phenomenal apparatus is but the consequence of the production of the carbonic acid and of its liberations—*i.e.*, the consequence of the act of destruction of the sugar. Here is organic destruction with its energetic manifestations!

This example of the life of brewers' yeast, of the saccharomyces, specially chosen by Le Dantec as being absolutely clear and giving the best illustration of his argument, contradicts him at every point. The general thesis of this vigorous thinker is that we cannot distinguish between the two parts of the vital act, organic destruction, and assimilating synthesis; that these two acts are not successive; that they give rise to phenomenal manifestations equally evident, apparent, or striking. Now, in the case of yeast, the phenomenon of destruction is clearly distinct from that of the assimilating synthesis which multiplies the substance of the saccharomyces. In fact, the action is realized by means of an alcoholic diastase manufactured by the cell; and Büchner succeeded in isolating this alcoholic ferment which splits up the sugar into alcohol and carbonic acid, and also *in vitro* and *in vivo*, makes the vat boil and heats the liquid. All the yeast is at work at once, says M. Dantec. No, and this is the proof.

And, further, Pasteur himself, who had shown the relation of the decomposition of the sugar to the fact of the growth of the yeast and of the production of accessory substances such as succinic acid and glycerine, always referred to *correlation* between these phenomena. The destruction of the sugar is the *correlative* of the life of the yeast. This was his favourite formula. It never entered his head that there could be a confusion instead of a correlation, and that there might be only one and the same act, the phases of which would be indistinguishable. This unfortunate idea, which was fated to be so rapidly contradicted, is due to Le Dantec. Far from it being the case, Pasteur had distinguished the *ferment function* from the life of the yeast. According to him, the yeast may exist sometimes as a ferment and sometimes otherwise.

§ 3. CORRELATION OF TWO ORDERS OF VITAL FACTS.

It is this correlation between acts *distinct in themselves* but *usually connected* that was announced by Claude Bernard. And, *mirabile dictu*—and this is the natural outcome of the perfect sanity of mind of this great physiologist—it happens that not only Pasteur's researches, but the development of a new science, Energetics, and Büchner's discovery lend support to his views, and that, too, in a field where one would have thought they had no application. Le Dantec is wrong when he declares that these ideas only apply to vertebrates. "It is clear," he says on several occasions, "that the author has in view the metazoa and even the vertebrates." Well! no. All that is general, universally applicable, and universally true. So that there are two orders of distinct phenomena energetically opposed and certainly connected. We need only repeat Claude Bernard's own words quoted by Le Dantec in order to confute them.

Law of Connection of Two Orders of Vital Facts.—"These phenomena [of organic destruction and of assimilating synthesis] are simultaneously produced in every living being, in a connection which cannot be broken. The disorganization or dissimilation uses up living matter [by this we must understand the reserve-stuff, as will be seen later on in the quotation] in the organs *in function:* the assimilating synthesis regenerates the tissues; it gathers together the reserve-stuff which the vital activity must expend. These two operations of destruction and renovation, inverse the one to the other, are absolutely connected and inseparable, in this sense at any rate, that destruction is the necessary condition of renovation. The phenomena of functional destruction are themselves the precursors and the instigators of material regeneration, of the formative process which is silently going on in the intimacy of the tissues. The losses are repaired as they take place; and equilibrium being re-established as soon as it tends to be broken, the body is maintained in its composition."

It is perfectly right and wise to say with Claude Bernard that the two orders of facts are successive, and that one is normally the inciting condition of the other. The possibility of the development of the yeast when fermentation fails, and the weakness of this development on the other hand under these conditions, are an excellent proof of this. The one proves the essential independence of the two orders of facts, the other the inciting and provoking virtue of the first relatively to the second. The experimental truth is thus expressed with a minimum of uncertainty. We know the facts which led Le Dantec to formulate his law of functional assimilation—namely, that the functional activity is useful or indispensable to the growth of the organ; that the organs which are functionally active grow, and those which do not act become atrophied. We are only expressing the facts when we say that the organic destructions that go on in the living being (whether at the expense of its reserve-stuff or at the expense of its medium, or whether it be even slightly at the expense of the plastic substance itself) are the antecedent, the inciting agent or the normal condition of the chemical and organogenic syntheses which create the new protoplasm.

On the other hand, we are wrong if we hold with Le Dantec that instead of two chemical operations there is only one, that which creates the new protoplasm. The obvious destruction is neglected; it is deliberately passed over. He does not see that it is necessary to liberate the energy employed in the construction, by complication, of this highly complex substance which is the new protoplasm. He really seems to have made up his mind not to analyze the phenomenon. If we decline to admit that to the first act of functional destruction succeeds a second, assimilation or organogenic synthesis, we are looking at elementary beings, in which the succession cannot be grasped, as we look on brewers' yeast. We not only mean that the morphogenic assimilation results from the functional activity; we mean that it results from it directly, immediately, that it is the functional activity itself. Experiment tells us nothing of all this. It shows us the real facts, the facts of the destruction of an organic immediate principle, the sugar, and the fact that an assimilating synthesis is the correlative of this destruction. Besides, if it is impossible in examples of this kind to exhibit the succession, it is perfectly easy in beings of a higher order. It is, then, clearly seen that the preliminary destruction of a reserve-stuff (and perhaps of a small quantity of the living substance) precedes and conditions the formation of a greater quantity of this living matter—in other words, the growth of the protoplasm of the organ.

Contradictions in the New Theory.—Moreover, these mistakes involve those who make them in a series of inextricable contradictions. Here, for example, is life; it is found, they say, in three forms:—Life manifested, or condition 1°; latent life, or condition 3°. So far this is the classical theory; but they add a condition 2°, which is what might be called *pathological or incomplete life.* This is defined by the following characteristic:—That its functional phenomena are identical with those in the first form, but that they are not accompanied by assimilation and by protoplasmic growth. But since, they say, growth is the chemical consequence of the functional activity, since it is so to speak its metabolic aspect, since it is confused with it, and inseparable from it, by the argument—then it is contradictory and logically absurd to speak of condition 2°. It would be acknowledging in the case of the anucleated merozoite, for example, a functional activity unaccompanied by assimilation, yet identical with the functional activity which is accompanied by assimilation in the nucleated merozoite. The general movement, that of the cilia, the taking of food, the evacuation of the fæces, the contraction of the pulsatile vacuoles, are the same. And this fact is the best proof that this vital functional activity (with the organic destruction which is its energetic source) must be distinguished from the assimilation which usually follows it, and which in exceptional cases may not follow it.

We shall carry this discussion no farther. We have examined at some length Le Dantec's views, and we have contrasted them with the doctrine which has been current in general physiology since the time of Claude Bernard, and this comparison does not turn out quite to their advantage. It was inevitable that the experimental and realistic spirit which inspired the doctrine of the celebrated physiologist made his work really too systematic. His formula, "life is death," and the form he gave his ideas, are not always irreproachably correct. They lend themselves at times to criticism. Sometimes they require commentary. These are errors of detail which Le Dantec has summarized somewhat roughly. There is no necessity to do this in his own case. We pay our tribute to the clearness of his language, although we believe the foundations of his system are false and ill-founded. Their rigour is purely verbal. Their external qualities, their careful arrangement are well adapted to the seduction of the systematic mind prepared by mathematical teaching. This new theory of life is presented with pedagogic talent of the highest order. We think we have shown that the foundations are entirely fallacious, in particular the following:—Vital condition No. 2°; the confusion between functional activity and assimilating synthesis; the so-called absolute connection between morphogeny and chemical composition; the fundamental distinction between elementary life and individual life.

§ 4. CHARACTERISTICS OF NUTRITION.

Definition of Nutrition.—As we have just seen, the organism is the scene of chemical reactions of two kinds, the one destructive and simplifying, the other synthetic, constructive, or assimilating. This totality of reactions constitutes nutrition. Hence the two phases that it is convenient to consider in this function—*assimilation* and *disassimilation*. This twofold chemical movement or *metabolism* corresponding to the two categories of vital phenomena, of destruction (catabolism) and of synthesis (anabolism) is therefore the chemical sign of vitality in all its forms. But it is clear that disassimilation or organic destruction, which is destined to furnish energy to the organism for its different operations, reappears in the plan of the general phenomena of nature. It is not specifically vital in its principle. Assimilation, on the other hand, is in this respect much more characteristic.

To some physiologists nutrition is only assimilation. Of the two aspects of metabolism they consider only one, the most typical, *Ad-similare*, to assimilate, to restore the substance borrowed from the ambient medium, the alimentary substances, *similar* to living matter, to make living matter of them, to increase active protoplasm—this is indeed the most striking phenomenon of vitality. To grow, to increase, to expand, to invade, is the law of living matter. Assimilation, nutrition in its essentials, is, according to the definition of Ch. Robin, "the production by the living being of a substance identical with its own." It is the act by which the living matter, the protoplasm of a given being, is created.

Permanence in Nutrition.—Nutrition presents one quite remarkable character—permanence. It is a vital manifestation, a property if we look at it in the cell, in the living substance, a function if we consider it in the animal or in the plant as a whole, which is never arrested. Its suspension involves *ipso facto* the suspension of life itself. It is, in the words of Claude Bernard, that property of nutrition "which, as long as it exists in an element, compels us to believe that this element is alive, and which, when it is absent, compels us to believe that it is dead. It dominates all others by its generality and its importance. In a word, it is the absolute test of vitality."

Biological Energetics shows the Importance of Nutrition.—We have indicated in advance the reason of its importance, showing that its two phases, dis-assimilation and assimilation, are the energetic condition of the two kinds of vital phenomena which we can distinguish.

Nutrition is a manufacture of protoplasm at the expense of the materials of the cellular ambient medium, which are assimilated—*i.e.*, made chemically and physically similar to living matter and to the reserves it stores up. This operation, which is peculiarly chemical, is therefore indicated by the borrowing of materials from the external world, a borrowing which is always going on, because the operation is permanent, and, let me add, because of the continual rejection of the waste products of this manufacture. Our formula is:—Nutrition is a chemistry which persists.

The Idea of the Vital Vortex is Erroneous.—Here the effect has hidden the cause from the eyes of the biologists. They have been struck by the incessant entry and exit, by the never-ceasing passage, by the *cycle* of matter through the living being without guessing its why and wherefore; and they have taken as a picture of the living being a vortex in which the essential form is maintained while the matter, which is accessory, flows on without a check. This is Cuvier's *vital vortex.* But for what purpose is this circulating matter used? They thought that it was employed entirely for the reconstitution of the living substance, continually and inevitably destroyed by the vital Minotaur.

Destruction of Reserve-stuff.—Here again there is a mistake. Really living substance is but little destroyed, and consequently requires very little renewal by the functional activity of the animal machine. Its metabolism—destruction and renewal—is in every case infinitely less than is supposed in the classical image of the vital vortex. It is the merit of physiologists, and particularly of Pflüger and Chauveau, to have worked for nearly forty years to establish this truth. They have proved it, at least as far as the muscular tissue is concerned. Protoplasm, properly so-called, is only destroyed as the organs of a steam engine are destroyed—its tubes, its boiler, its furnace. And it matters little. We know that such an engine uses much coal, and we know very little of its machinery and its metallic frame. And so it is with the cell, the living machine. A very small portion of the food introduced will be assimilated in the living substance. By far the greater part of it is destined to be worked up by the protoplasm and placed in reserve under the form of glycogen, albumen, and fat, etc.—*i.e.*, compounds which are not the really living substance, the hereditary protoplasm, but the products of its industry, just as they are or may be the products of the industry of the chemist working in his laboratory. They will be expended for the purpose of furnishing the necessary energy to the vital functional activity, muscular contraction, secretion, heat, etc., just as coal is expended to set the steam engine going. The proof as far as the muscle is concerned does not stand alone. There are other examples. In particular, micrographic physiologists who have studied nervous phenomena say that the anatomical elements of the brain last indefinitely, and that they continue as they are, without renewal from birth to death. The permanence of the consciousness, be it said in passing, is connected by them with the permanence of the cerebral element (Marinesco).

Thus destruction is very restricted. There is only a very slight disassimilation of the living matter, properly so-called, in the course of the vital functional activity. We may even go farther than this experimental fact. This is what Le Dantec has done when he claims that there is even an assimilation, an increase of the protoplasm. Strictly speaking, this is possible, but there is no certain proof of it; and in any case we cannot agree with him when he affirms that the increase is the *direct result* of the functional activity and blends with it in one single, unique operation. We must, on the contrary, agree with Claude Bernard that it is only a *consequence* of it, that it is produced in consequence of the existence of a bond of correlation between organic destruction and assimilating synthesis.

Why is there this bond? That is easily understood if we reflect that the assimilating synthesis, an operation of endothermic, chemical complexity, naturally requires an exothermic counterpart, the organic destruction which will set free this necessary energy.

Formative Assimilation of Reserve-stuff. Formative Assimilation of Protoplasm.—It follows that there are in nutritive assimilation itself two distinct acts. The one consisting of the manufacture of reserve-stuff is the more obvious but the less specific; the other, really essential, is assimilation properly so-called, the reconstitution of the protoplasm. The former is indispensable to the production of the most prominent acts of vitality—movement, secretion, production of heat. If it is suspended, functional activity is arrested. We get *apparent death*, or *latent life*. But if the real assimilation is arrested, we have *real death*.

According to this there would be a fundamental distinction between real and apparent death. The former would be characterized by an *arrest of the protoplasmic assimilation* which is externally indicated by no sign. On the other hand, apparent death would be characterized by *the arrest of the formation and destruction of reserve-stuff*. It would be externally manifested by two signs:—The suppression of material exchanges with the medium (respiration, alimentation) and the suppression of the functional acts (production of movement, of heat, of electricity, of glandular excretion).

Such would be the most expedient test for apparent or real death. The question occurs in the case of grains of corn in Egyptian tombs, and also of hibernating animals and reviviscent beings, and, in general, in the case of what has been called the state of *latent life*. But from the practical point of view it is extremely difficult to apply this test and to decide if the phenomena which are arrested in the grain at maturity, in Leeuwenhoek's tardigrada,[1]

[1] Bear-animalcules, Sloth-animalcules. An order of Arachnida.—Tr.

and in the dried-up Anguillulidæ[1] of Baker and Spallanzani, in the encysted colpoda[2] that a drop of warm water will revive, in the animals exposed by E. Yung and Pictet to a cold of more than a 100° C. below zero, are due to the general arrest of the two forms of assimilation, or to the arrest of the manufacture and utilization of reserve-stuff alone, or finally, to the arrest of protoplasmic assimilation alone. The latter, which is already very restricted in beings in a normal condition whose growth is terminated, may fall to the lowest degree in the being which, having no functional activity, is assimilating nothing. So that, to cut the question short, the experimenter who measures the value of the exchanges between the being and the medium has seldom to do more than decide between little and nothing. Hence his perplexity. But if experiment hesitates, theory affirms: it admits *a priori* that the movement of protoplasmic assimilation, an essential sign of vitality, is neither checked nor renewed, but proceeds continuously.

Is Nutrition, the Assimilating Synthesis, interrupted?—Nevertheless, there are many reasons for suspending all judgment as to this interpretation. It is questioned by most biologists. According to A. Gautier, the preserved grain of corn and the dried up rotifera are not really alive; they are like clocks in working order, ready to tell the time, but awaiting in absolute repose the first vibration which will set them going. As for the grain, it is the air, heat, and

[1] Minute thread worms, known as paste-eels and vinegar-eels.—Tr.

[2] Genus of Infusoria. *Colpodea cucullus* is found in infusions of hay—Tr.

moisture which supply the first impulse. In other words, the organization proper to the manifestation of life remains, but there is no life. The so-called arrested life is not a life.

It must be said, however, that the majority of physiologists refuse to accept this interpretation. They believe in an attenuation of the nutritive synthesis and not in its complete destruction. They think that this total suppression would be contrary to current ideas relative to the perpetuity of the protoplasm and the limited duration of the living element. The natural medium is variable, and even the mineral cannot remain eternally fixed. Still less is perennity a property of the living being. If ordinary life is for each individual of limited duration, the arrested life must also be of limited duration. We cannot believe that after an indefinitely prolonged sleep the grain of corn, or the paste-eel, or the colpoda, emerging from their torpor can resume their existence, like the Sleeping Beauty, at the point at which it was interrupted, and thus pass with a bound, as it were, through the centuries.

In fact the maintenance of the vitality of grains of corn from the Egyptian tombs and their aptitude to germinate after thousands of years are only fables or the result of imposture. Maspero, in a letter addressed to M. E. Griffon on the 15th July 1901, has clearly summed up the situation by saying that the grains of corn bought from the fellahs almost always germinate, but that this is never the case with those that the experimenter himself takes from the tombs.

To sum up, we must use the same language of nutrition and of life, of their uninterrupted progress, of their continuity, of their permanence, of their activity, and of their slackening. Living matter is always growing, much or little, slowly or quickly, in its reserve-stuff or in its protoplasm, for expenditure or accumulation. This inevitability of growth defines it, characterizes it, and sums up its activity. Development and the evolution of growth are consequences or aspects of nutrition.

BOOK IV.

THE LIFE OF MATTER.

Apparent Differences between Living and Brute Bodies. The Two Kingdoms.—It seems at first impossible that there should be any essential similarity between an inanimate object and a living being. What resemblance can be discovered between a stone, a lion, and an oak? A comparison of the inert and immovable pebble with the leaping animal, and with the plant extending its foliage gives an impression of vivid contrast. Between the organic and the inorganic worlds there seems to be an abyss. The first impressions we receive confirm this view; superficial investigation furnishes arguments for it. There is thus aroused in the mind of the child, and later in that of the man, a sharply marked distinction between the natural objects of the mineral kingdom on the one hand, and those of the two kingdoms of living beings on the other.

But a more intimate knowledge daily tends to throw doubt upon the rigour or the absolute character of such a distinction. It shows that brute matter can no longer be placed on one side and living beings on the other. Scientists deliberately speak of "the life of matter," which seems to the average man a contradiction in terms. They discover in certain classes of mineral bodies almost all the attributes of life. They find in others fainter, but still recognizable indications of an undeniable relationship.

We propose to pass in review these analogies and resemblances, as has already been done in a fairly complete manner by Leo Errera, C. E. Guillaume, L. Bourdeau, Ed. Griffon, and others. We will consider the fine researches of Rauber, of Ostwald, and of Tammann upon crystals and crystalline germs —researches which are merely a continuation of those of Pasteur and of Gernez. These show that crystalline bodies are endowed with the principal attributes of living beings—*i.e.*, a rigorously defined form; an aptitude for acquiring it, and for re-establishing it by repairing any mutilations that may be inflicted upon it; nutritive growth at the expense of the mother liquor which constitutes its culture medium; and, finally—a still more incredible property—all the characteristics of reproduction by generation. Other curious facts observed by skilful physicists—W. Roberts-Austen, W. Spring, Stead, Osmond, Guillemin, Charpy, C. E. Guillaume—show that the immutability even of bodies supposed to be the most rigid of all, such as glass, the metals, steel, and brass, is apparent rather than real. Beneath the surface of the metal that seems to us inert there is a swarming population of molecules, displacing each other, moving about, and arranging themselves so as to form definite figures, and assuming forms adapted to the conditions of the environment. Sometimes it is years before they arrive at the state of ultimate and final equilibrium—which is that of eternal rest.

However, in order to understand these facts and their interpretations, it is necessary to pass in review the fundamental characteristics of living beings. It will be shown that these very characteristics are found in inanimate matter.

CHAPTER I.

UNIVERSAL LIFE. OPINIONS OF PHILOSOPHERS AND POETS.

§ 1. Primitive beliefs; the ideas of poets.—§ 2. Opinions of philosophers—Transition from brute to living bodies—The principle of continuity: continuity by transition: continuity by summation—Ideas of philosophers as to sensibility and consciousness in brute bodies—The general principle of homogeneity—The principle of continuity as a consequence of the principle of homogeneity.

§ 1. PRIMITIVE BELIEFS. IDEAS OF THE POETS.

The teaching of science as to the analogies between brute bodies and living bodies accords with the conceptions of the philosophers and the fancies of the poets. The ancients held that all bodies in nature were the constituent parts of a universal organism, the macrocosm, which they compared to the human microcosm. They attributed to it a principle of action, the *psyche*, analogous to the vital principle, and this psyche directed phenomena; and also an intelligent principle, the *nous*, analogous to the soul, and the *nous* served for the comprehension of phenomena. This universal life and this universal soul played an important part in their metaphysical systems.

It was the same with the poets. Their tendency has always been to attribute life to Nature, so as to bring her into harmony with our thoughts and feelings. They seek to discover the life or soul hidden in the background of things.

> "Hark to the voices. Nothing is silent.
>
> Winds, waves, and flames, trees, reeds, and rocks
> All live; all are instinct with soul."

After making proper allowance for emotional exaggeration, ought we to consider these ideas as the prophetic divination of a truth which science is only just beginning to dimly perceive? By no means. As Renan has said, this universal animism, instead of being a product of refined reflection, is merely a legacy from the most primitive of mental processes, a residue of conceptions belonging to the childhood of humanity. It recalls the time when men conceived of external things only in terms of themselves; when they pictured each object of nature as a living being. Thus, they personified the sky, the earth, the sea, the mountains, the rivers, the fountains, and the fields. They likened to animate voices the murmur of the forest:—

> ". . . The oak chides and the birch
> Is whispering. . . .
> And the beech murmurs. . . .
> The willow's shiver, soft and faint, sounds like a word.
> The pine-tree utters mysterious moans."

For primitive man, as for the poet of all times, everything is alive, and every sound is due to a being with feelings similar to our own. The sighing of the breeze, the moan of the wave upon the shore, the babbling of the brook, the roaring of the sea, and the pealing of the thunder are nothing less than sad, joyous, or angry living voices.

These impressions were embodied in ancient mythology, the graceful beauty of which does not conceal its inadequacy. Then they passed into philosophy and approached the realm of science. Thales believed that all bodies in nature were animate and living. Origen considered the stars as actual beings. Even Kepler himself attributed to the celestial bodies an internal principle of action, which, it may be said in passing, is contrary to the law of the inertia of matter, which has been wrongly ascribed to him instead of to Galileo. The terrestrial globe was, according to him, a huge animal, sensitive to astral influences, frightened at the approach of the other planets, and manifesting its terror by tempests, hurricanes, and earthquakes. The wonderful flux and reflux of the ocean was its breathing. The earth had its blood, its perspiration, its excretions; it also had its foods, among which was the sea water which it absorbs by numerous channels. It is only fair to add that at the end of his life Kepler retracted these vague dreams, ascribing them to the influence of J. C. Scaliger. He explained that by the soul of the celestial bodies he meant nothing more than their motive force.

§ 2. Opinion of the Philosophers.

Transition from Brute to Living Bodies.—The lowering of the barrier between brute bodies and living bodies began with those philosophers who introduced into the world the great principles of continuity and evolution.

The Principle of Continuity.—First and foremost we must mention Leibniz. According to the teaching of that illustrious philosopher, as interpreted by M. Fouillée, "there is no inorganic kingdom, only a great organic kingdom, of which mineral, vegetable, and animal forms are the various developments. . . . Continuity exists everywhere throughout the world; everywhere is life and organization. Nothing is dead; life is universal." It follows that there is no interruption or break in the succession of natural phenomena; that everything is gradually developed; and finally, that the origin of the organic being must be sought in the inorganic. Life, properly so called, has not, in fact, always existed on the surface of the globe. It appeared at a certain geological epoch, in a purely inorganic medium, by reason of favourable conditions. The doctrine of continuity compels us, however, to admit that it pre-existed on the globe under some rudimentary form.

The modern philosophers who are imbued with these principles, MM. Fouillée, L. Bourdeau, and A. Sabatier, express themselves in similar language. "Dead matter and living matter are not two absolutely different entities, but represent two forms of the same matter, differing only in degree, sometimes but slightly." When it is only a matter of degree, it cannot be held that these views are opposed. Inequalities must not be interpreted as contrary attributes, as when the untrained mind considers heat and cold as objective states, qualitatively opposed to each other.

Continuity by Transition.—The argument which

leads us to remove the barrier between the two kingdoms, and to consider minerals as endowed with a sort of rudimentary life, is the same as that which compels us to admit that there is no fundamental difference between natural phenomena. There are transitions between what lives and what does not, between the animate being and the brute body. And in the same way there are transitions between what thinks and what does not think, between what is thought and what is not thought, between the conscious and the unconscious. This idea of insensible transition, of a continuous path between apparent antitheses, at first arouses an insuperable opposition in minds not prepared for it by a long comparison of facts. It is slowly realized, and finally is accepted by those who, in the world of things, follow the infinity of gradations presented by natural phenomena. The principle of continuity comes at last to constitute, as one may say, a mental habit. Thus the man of science may be led, like the philosopher, to entertain the idea of a rudimentary form of life animating matter. He may, like the philosopher, be guided by this idea; he may attribute *a priori* to brute matter all the really essential qualities of living beings. But this must be on the condition that, assuming these properties to be common, he must afterwards demonstrate them by means of observation and experiment. He must show that molecules and atoms, far from being inert and dead masses, are in reality active elements, endowed with a kind of inferior life, which is manifested by all the transformations observed in brute matter, by attractions and repulsions, by movements in response to external stimuli, by variations of state and of equilibrium; and finally, by the systematic methods according to which these elements group themselves, conforming to those definite types of structure by means of which they produce different species of chemical compounds.

Continuity by Summation.—The idea of summation leads by another path to the same result. It is another form of the principle of continuity. A sum total of effects, obscure and indistinct in themselves, produces a phenomenon appreciable, perceptible, and distinct, apparently, but not really, heterogeneous in its components. The manifestations of atomic or molecular activity thus become manifestations of vital activity.

This is another consequence of the teaching of Leibniz. For, according to his philosophical theory, individual consciousness, like individual life, is the collective expression of a multitude of elementary lives or consciousnesses. These elements are inappreciable because of their low degree, and the real phenomenon is found in the sum, or rather the *integral*, of all these insensible effects. The elementary consciousnesses are harmonized, unified, integrated into a result that becomes manifest, just as "the sounds of the waves, not one of which would be heard if by itself, yet, when united together and perceived at the same instant, become the resounding voice of the ocean."

Ideas of the Philosophers as to Sensibility and Consciousness in Brute Bodies.—The philosophers have gone still further in the way of analogies, and have recognized in the play of the forces of brute matter, particularly in the play of chemical forces, a mere rudiment of the appetitions and tendencies that regulate, as they believe, the functional activity of living beings—a trace, as it were, of their sensibility. To them reactions of matter indicate the existence of a kind of *hedonic consciousness—i.e.*, a consciousness reduced simply to a distinction between comfort and discomfort, a desire for good and repulsion from evil, which they suppose to be the universal principle of all activity. This was the view held by Empedocles in antiquity; it was that of Diderot, of Cabanis, and, in general, of the modern materialistic school, eager to find, even in the lowest representatives of the inorganic world, the first traces of the vitality and intellectual life which blossom out at the top of the scale in the living world.

Similar ideas are clearly seen in the early history of all natural sciences. It was this same principle of appetition, or of love and of repulsion or hate that, under the names of affinity, selection, and incompatibility, was thought to direct the transformations of bodies when chemistry first began; when Boerhaave, for example, compared chemical combinations to voluntary and conscious alliances, in which the respective elements, drawn together by sympathy, contracted appropriate marriages.

General Principle of the Homogeneity of the Complex and its Components.—The assimilation of brute bodies to living bodies, and of the inorganic kingdom to the organic, was, in the mind of these philosophers, the natural consequence of positing *a priori* the principles of continuity and evolution. There is, however, a principle underlying these principles. This principle is not expressed explicitly by the philosophers; it is not formulated in precise terms, but is more or less unconsciously implied; it is everywhere applied. It, however, may be clearly seen behind the apparatus of philosophical argument. It is the assertion that no arrangement or combination of elements can put forth any new activity essentially different from the activities of the elements of which it is composed. Man is living clay, say Diderot and Cabanis; and, on the other hand, he is a thinking being. *As it is impossible to produce that which thinks from that which does not think*, the clay must possess a rudiment of thought. But is there not another alternative? May not the new phenomenon, thought, be the effect of the arrangement of this clay? If we exclude this alternative, we must then consider arrangement and organization as incapable of producing in arranged and organized matter a new property different from that which it presented before such arrangement. Living protoplasm, says another, is merely an assemblage of brute elements; "these brute elements must therefore possess a rudiment of life." This is the same implied supposition which we have just considered; if life is not the basis of each element, it cannot result from their simple assemblage.

Man and animals are combinations of atoms, says M. le Dantec. It is more natural to admit that human consciousness is the result of the elementary consciousness of the constituent atoms than to consider it as resulting from construction by means of elements with no consciousness. "Life," says Haeckel, "is universal; we could not conceive of its existence in certain aggregates of matter if it did not belong to their constituent elements." Here the postulate is almost expressed.

The argument is always the same; even the same words are used: the fundamental hypothesis is the same; only it remains more or less unexpressed, more or less unperceived. It may be stated as follows:—Arrangement, assemblage, construction, and aggregation are powerless to bring to light in the complex anything new and essentially heterogeneous to what already exists in the elements. Reciprocally, grouping reveals in a complex a property and character which is the gradual development of an analogous property and character in the elements. It is in this sense that there exists a collective soul in crowds, the psychology of which has been discussed by M. G. Le Bon. In the same way, many sociologists, adopting the views advanced by P. de Lilienfeld in 1865, attribute to nations a formal individuality, after the type of that possessed by each of their constituent members. M. Izolet considers society as an organism, which he calls a "hyperzoan." Herbert Spencer has developed the comparison of the collective organism with the individual organism, insisting on their resemblances and differences. Th. Ribot has dwelt, in particular, on the resemblances.

The postulate that we have clearly stated here is accepted by many as an axiom. But it is not an axiom. When we say that there is nothing in the complex that cannot be found in the parts, we think we are expressing a self-evident truth; but we are, in fact, merely stating an hypothesis. It is assumed that arrangement, aggregation, and complicated and skilful grouping of elements can produce nothing really new in the order of phenomena. And this is an assertion that requires verification in each particular case.

The Principle of Continuity, a Consequence of the Preceding.—Let us apply this principle to the beings in nature. All beings in nature are, according to current ideas, arrangements, aggregates, or groupings of the same universal matter, that is to say, of the same simple chemical bodies. It results from the preceding postulate that their activities can only differ in degree and form, and not fundamentally. There is no essential difference of nature between the activities of various categories of beings, no heterogeneity, no discontinuity. We may pass from one to another without coming to an hiatus or impassable gulf. The law of continuity thus appears as a simple consequence of the fundamental postulate. And so it is with the law of evolution, for evolution is merely continuity of action.

Such are the origins of the philosophical doctrine which universalizes life and extends it to all bodies in nature.

It may be remarked that this doctrine is not confined to any particular school or sect. Leibniz was by no means a materialist, and he endowed his mundane elements, his *monads*, not only with a sort of life, but even with a sort of soul. Father Boscovitch, Jesuit as he was, and professor in the college of Rome, did not deny to his *indivisible points* a kind of inferior vitality. St. Thomas, too, the angelical doctor, attributed, according to M. Gardair, to inanimate substances a certain kind of activity, inborn inclinations, and a real appetition towards certain acts.

CHAPTER II.

ORIGIN OF BRUTE MATTER IN LIVING MATTER.

Spontaneous generation: an episode in the history of the globe—Verification of the identity between brute and living matter—Slow identification—Rapid identification—Contrary opinion—Hypothesis of cosmozoa; cosmic panspermia—Hypothesis of pyrozoa.

THERE should be two ways of testing the doctrine of the essential identity of brute and living matter—one slow and more laborious, the other more rapid and decisive.

Identification of the Two Matters, Brute and Living.—The laborious method, which we will be obliged to follow, consists in the attentive examination of the various activities by which life is manifested, and in finding more or less crude equivalents for them in all brute beings, or in certain of them.

Rapid Verification. Spontaneous Generation.—The rapid and decisive method, which, unhappily, is beyond our resources, would consist in showing unquestionable, clearly marked life, the superior life, arising from the kind of inferior life that is attributed to matter in general. It would be necessary completely to construct in all its parts, by a suitable combination of inorganic materials, a single living being, even the humblest plant or the most rudimentary animal. This would indeed be an irrefutable proof that the germs of all vital activity are contained in the molecular activity of brute bodies, and that there is nothing essential to the latter that is not found in the former.

Unhappily this demonstration cannot be given. Science furnishes no example of it, and we are forced to have recourse to the slow method.

The question here involved is that of spontaneous generation. It is well known that the ancients believed in spontaneous generation, even for animals high in the scale of organization. According to Van Helmont, mice could be born by some incomprehensible fermentation in dirty linen mixed with wheat. Diodorus speaks of animal forms which were seen to emerge, partly developed, from the mud of the Nile. Aristotle believed in the spontaneous birth of certain fishes. This belief, though rejected as to the higher forms, was for a long time held with regard to the lower forms of animals, and to insects—such as the bees which the shepherd of Virgil saw coming out from the flanks of the dead bullock—flies engendered in putrefying meat, fruit worms and intestinal worms; finally, with regard to infusoria and the most rudimentary vegetables. The hypothesis of the spontaneous generation of the living being at the expense of the materials of the ambient medium has been successively driven from one classificatory group to another. The history of the sciences of observation is also a history of the confutation of this theory. Pasteur gave it the finishing stroke, when he showed that the simplest micro-organisms obeyed the general law which declares that the living being is formed only by *filiation*—that is to say, by the intervention of a pre-existing living organism.

Spontaneous Generation an Episode in the History of the Globe.—Though we have been unable to effect spontaneous generation up to the present, it has been referred by Haeckel to a more or less distant past, to the time when the cooling of the globe, the solidification of its crust, and the condensation of aqueous vapour upon its surface created conditions compatible with the existence of living beings similar to those with which we are acquainted. Lord Kelvin has fixed these geological events as occurring from twenty to forty million years ago. Then circumstances became propitious for the appearance of the first organisms, whence were successively derived those which now people the earth and the waters.

Circumstances favourable to the appearance of the first beings apparently occurred only in a far distant past; but most physiologists admit that if we knew exactly these circumstances, and could reproduce them, we might also expect to produce their effect—namely, the creation of a living being, formed in all its parts, developed from the inorganic kingdom. To all those who held this view the impotence of experiment at the present time is purely temporary. It is comparable to that of primitive men before the time of Prometheus; they, not knowing how to produce fire, could only get it by transmitting it from one to another. It is due to the inadequacy of our knowledge and the weakness of our means; it does not contradict the possibility of the fact.

Contrary Opinion. Life did not Originate on our Globe.—But all biologists do not share this opinion. Some, and not the least eminent, hold it to be an established fact that it is impossible for life to arise from a concurrence of inorganic materials and forces. This was the opinion of Ferdinand Cohn, the great botanist; of H. Richter, the Saxon physician, and of W. Preyer, a physiologist well known from his remarkable researches in biological chemistry. According to these scientists, life on the surface of the globe cannot have appeared as a result of the reactions of brute matter and the forces that continue to control it.

According to F. Cohn and H. Richter, life had no beginning on our planet. It was transported to the earth from another world, from the cosmic medium, under the form of cosmic germs, or *cosmozoa*, more or less comparable to the living cells with which we are acquainted. They may have made the journey either enclosed in meteorites, or floating in space in the form of cosmic dust. The theory in question has been presented in two forms:—*The Hypothesis of Meteoric Cosmozoa*, by a French writer, the Count de Salles-Guyon; and that of *cosmic panspermia* brought forward in 1865 and 1872 by F. Cohn and H. Richter.

Hypothesis of the Cosmozoa.—The hypothesis of the *cosmozoa*, living particles, protoplasmic germs emanating from other worlds and reaching the earth by means of aerolites, is not so destitute of probability as one might at first suppose. Lord Kelvin and Helmholtz gave it the support of their high authority. Spectrum analysis shows in cometary nebulæ the four or five lines characteristic of hydro-carbons. Cosmic matter, therefore, contains compounds of carbon, substances that are especially typical of organic chemistry. Besides, carbon and a sort of humus have been found in several meteorites. To the objection that these aerolites are heated while passing through our atmosphere, Helmholtz replies that this elevation of temperature may be quite superficial and may allow micro-organisms to subsist in their interior. But other objections retain their force:—First, that of M. Verworn, who considers the hypothesis of cosmic germs as inconsistent with the laws of evolution; and that of L. Errera, who denies that the conditions necessary for life exist in inter-planetary bodies.

Hypothesis of Cosmic Panspermia.—Du Bois-Reymond has given the name of *cosmic panspermia* to a theory very similar to the preceding, formulated by F. Cohn in 1872. The first living germs arrived on our globe mingled with the cosmic dust that floats in space and falls slowly to the surface of the earth. L. Errera observes that if they escape by this gentle fall the dangerous heating of meteorites, they still remain exposed to the action of the photic rays, which is generally destructive to germs.

Hypothesis of Pyrozoa.—W. Preyer declined to accept this cosmic transmigration of the simplest living beings, nor would he allow the intervention of other worlds into the history of our own. Life, according to him, must have existed from all time, even when the globe was an incandescent mass. But it was not the same life as at present. Vitality must have undergone many profound changes in the course of ages. The *pyrozoa*, the first living beings, vulcanians, were very different from the beings of the present day that are destroyed by a slight elevation of temperature. No doubt this theory of pyrozoa, proposed by W. Preyer in 1872, seems quite chimerical, and akin to Kepler's dreamy visions. But in a certain way it accords with contemporary ideas concerning the life of *matter*. It is related to them by the evolution which it implies in the materials of the terrestrial globe.

According to Preyer, primitive life existed in fire. Being igneous masses in fusion, the pyrozoa lived after their own manner; their vitality, slowly modified, assumed the form which it presents to-day. Yet, in this profound transformation their number has not varied, and the total quantity of life in the universe has remained unchanged.

Here we recognize the ideas of Buffon. These cosmozoa, these pyrozoa, have a singular resemblance to the *organic molecules* of "live matter" of the illustrious naturalist—distributed everywhere, indestructible, and forming living structures by their concentration.

But we must leave these scientific or philosophical theories, and come to arguments based upon facts.

It is in a spirit quite different from that of the poets, the metaphysicians, and the more or less philosophical scientists that the science of our days looks at the more or less obscure vitality of inanimate bodies. It claims that we may recognize in them, in a more or less rudimentary state, the action of the factors which intervene in the case of living beings, the manifestation of the same fundamental properties.

CHAPTER III.

ORGANIZATION AND CHEMICAL COMPOSITION OF LIVING AND BRUTE MATTER.

Laws of the organization and of the chemical composition of living beings—Relative value of these laws; vital phenomena in crushed protoplasm—Vital phenomena in brute bodies.

Enumeration of the Principal Characters of Living Beings.—The programme which we have just sketched compels us to look in the brute being for the properties of living beings. What, then, are, in fact, the characteristics of an authentic, complete, living being? What are its fundamental properties? We have enumerated them above as follows:—A certain chemical composition, which is that of living matter; a structure or organization; a specific form; an evolution which has a duration, that of life, and an end, death; a property of growth or nutrition; a property of reproduction. Which of these characters counts for most in the definition of life? Are they all equally necessary? If some of them were wanting, would that justify the transference of a being, who might possess the rest, from the animate world to that of minerals? This is precisely the question that is under consideration.

Organization and Chemical Composition of Living Beings.—All that we know concerning the constitution of living matter and its organization is summed up in the laws of the *chemical unity* and the *morphological unity of living beings* (v. Book III.). These laws seem to be a legitimate generalization from all the facts observed. The first states that the phenomena of life are manifested only in and through living matter, protoplasm—*i.e.*, in and through a substance which has a certain chemical and physical composition. Chemically it is a proteid complexus with a hexonic nucleus. Physically it shows a frothy structure analogous to that resulting from the mixture of two granular, immiscible liquids, of different viscosities. The second law states that the phenomena of life can only be maintained in a protoplasm which has the organization of the complete cell, with its cellular body and nucleus.

Relative Value of these Laws. Exceptions.—What is the signification of these laws of the chemical composition and organization of living beings? Evidently that life in all its plenitude can only exist and be perpetuated under their protection. If these laws were absolute, if it were true that no life were possible but in and through albuminous protoplasm, but in and through the cell, the problem of "the life of matter" would be decided in the negative.

May it not happen, however, that fragmentary and incomplete vital manifestations, progressive traces of a true life, may occur under different conditions; for example, in matter which is not protoplasm, and in a body which has a structure differing from that of a cell—that is to say, in a being which would be neither animal nor plant? We must seek the answer to this question by an appeal to experiment.

Without leaving the animal and vegetable kingdoms—*i.e.*, real living beings—we already see less rigour in the laws governing chemical constitution and cellular organization.

Experiments in merotomy—*i.e.*, in amputation—carried out on the nervous element by Waller, on infusoria by Brandt, Gruber, Balbiani, Nussbaum, and Verworn, show us the necessity of the presence of the cellular body and the nucleus—*i.e.*, of the integrity of the cell. But they also teach us that when that integrity no longer exists death does not immediately follow. A part of the vital functions continues to be performed in denucleated protoplasm, in a cell which is mutilated and incomplete.

Vital Phenomena in Crushed Protoplasm.—It is true also that grinding and crushing suppress the greater part of the functions of the cell. But tests with pulps of various organs and with those of certain yeasts also show that protoplasm, even though ground and disorganized, cannot be considered as inert, and that it still exhibits many of its characteristic phenomena; for example, the production of diastases, the specific agents of vital chemistry. Finally, while we do not know enough about the actions of which the secondary elements of protoplasm—its granulations, its filaments—are capable, which this or that method of destruction may bring to light, at least we know that actions of this kind exist.

To sum up, we are far from being able to deny that rudimentary, isolated vital acts may be produced by the various bodies that result from the dismemberment of protoplasm. The integrity of the cellular organization, even the integrity of protoplasm itself, are therefore not indispensable for these partial manifestations of vitality.

Besides, biologists admit that there exist within the protoplasm aliquot parts, elements of an inferior order, which possess special activities. These secondary elements must have the principle of their activity within themselves. Such are the *biophors* to which Weismann attributes the vital functions of the cell, nutrition, growth, and multiplication. If there are biophors within the cell, we may imagine them outside the cell, and since they carry within themselves the principle of their activity they may exercise it in an independent manner. Unhappily the biophors, and other constituent elements of that kind, are purely hypothetical. They are like Darwin's gemmules, Altmann's bioblasts, and the pangens of De Vries. They have no relation to facts of observation and to real existence.

Vital Phenomena in Brute Bodies.—There is no doubt that certain phenomena of vitality may occur outside of the cellular atmosphere. And carrying this further, we may admit that they may be produced in certain slightly organized bodies (crushed cells), and then in certain unorganized bodies in certain brute beings. In every case it is certain that effects are produced at any rate similar to those which are characteristic of living matter. It is for observation and experiment to decide as to the degree of similarity, and their verdict is that the similarity is complete. The crystals and the crystalline germs studied by Ostwald and Tammann are the seat of phenomena which are quite comparable to those of vitality.

CHAPTER IV.

EVOLUTION AND MUTABILITY OF LIVING MATTER AND BRUTE MATTER.

Supposed immobility of brute bodies—Mobility and mutability of the sidereal world.—§ 1. The movement of particles and molecules in brute bodies—The internal movements of brute bodies—Kinetic conception of molecular motion—Reality of the motion of particles—Comparison of the activity of particles with vital activity.—§ 2. Brownian movement—Its existence—Its character—Its independence of the nature of the bodies and of the nature of the environment—Its indefinite duration—Its independence of external conditions—The Brownian movement must be the first stage of molecular motion.—§ 3. Motion of particles—Migration of material particles — Migration under the action of weight; of diffusion; of electrolysis; of mechanical pressure.—§ 4. Internal activity of alloys—Their structure—Changes produced by deforming agencies—Slow return to equilibrium—Residual effect—Effect of annealing; effect of stretching—Nickel steel—Colour photography—Conclusion—Relations of the environment to the living or brute matter.

ONE of the most remarkable characteristics of a living being is its evolution. It undergoes a continuous change. It starts from something very small; it assumes a configuration and grows; in most cases it declines and disappears, having followed a course which may be predicted—a sort of ideal trajectory.

Supposed Immobility of Brute Bodies.—It may be asked whether this evolution, this directed mobility, is so exclusively a feature of the living being as it appears, and if many brute bodies do not present something analogous to it. We may answer in no uncertain tones.

Bichat was wrong when he contrasted in this respect brute bodies with living bodies. Vital properties, he said, are temporary; it is their nature to be exhausted; in time they are used up in the same body. Physical properties, on the contrary, are eternal. Brute bodies have neither a beginning nor an inevitable end, neither age, nor evolution; they remain as immutable as death, of which they are the image.

Mobility and Mutability of the Sidereal World.—This is not true, in the first place, of the sidereal bodies. The ancients held the sidereal world to be immutable and incorruptible. The doctrine of the incorruptibility of the heavens prevailed up to the seventeenth century. The observers who at that epoch directed towards the heavens the first telescope, which Galileo had just invented, were struck with astonishment at discovering a change in that celestial firmament which they had hitherto believed incorruptible, and at perceiving a new star that appeared in the constellation Ophiúchus. Such changes no longer surprise us. The cosmogonic system of Laplace has become familiar to all cultivated minds, and every one is accustomed to the idea of the continual mobility and evolution of the celestial world. "The stars have not always existed," writes M. Faye; "they have had a period of formation; they will likewise have a period of decline, followed by final extinction."

Thus all the bodies of inanimate nature are not

eternal and immutable; the celestial bodies are eminently susceptible of evolution, slow indeed with that we observe on the surface of our globe; but this disproportion, corresponding to the immensity of time and of cosmic spaces as compared with terrestrial measurements, should not mislead us as to the fundamental analogy of the phenomena.

§ I. The Movement of Particles and Molecules in Brute Bodies.

It is not only in celestial spaces that we must search for that mobility of brute matter which imitates the mobility of living matter. In order to find it we have only to look about us, or to inquire from physicists and chemists.

As far as geologists are concerned, M. le Dantec tells us somewhere of one who divided minerals into *living rocks*—rocks capable of change of structure, of evolution under the influence of atmospheric causes; and *dead rocks*—rocks which, like clay, have found at the end of all their changes a final state of repose. Jerome Cardan, a celebrated scientist of the sixteenth century, at once mathematician, naturalist, and physician, declared not only that stones live, but that they suffer from disease, grow old, and die. The jewellers of the present day use similar language of certain precious stones; the torquoise, for example.

The alchemists carried these ideas to an extreme. It is not necessary here to recall the past, to evoke the hermetic beliefs and the dreams of the alchemists, who held that the different kinds of matter lived, developed, and were transmuted into each other.

I refer to precise and recent data, established by the most expert investigators, and related by one of them, Charles Edward Guillaume, some years ago, before the *Société helvétique des Sciences naturelles.* These data show that determinate forms of matter may live and die, in the sense that they may be slowly and continuously modified, always in the same direction, until they have attained an ultimate and definitive state of eternal repose.

The Internal Movements of Bodies.—Swift's reply to an idle fellow who spoke slightingly of work is well known. "In England," said the author of *Gulliver's Travels,* "men work, women work, horses work, oxen work, water works, fire works, and beer works; it is only the pig who does nothing at all; he must, therefore, be the only gentleman in England." We know very well that English gentlemen also work. Indeed, everybody and everything works. And the great wit was nearer right than he supposed in comparing men and things in this respect. Everything is at work; everything in nature strives and toils, at every stage, in every degree. Immobility and repose in the case of natural things are usually deceptive; the seeming quietude of matter is caused by our inability to appreciate its internal movements. Because of their minuteness we do not perceive the swarming particles that compose it, and which, under the impassible surface of the bodies, oscillate, displace each other, move to and fro, and group themselves into forms and positions adapted to the conditions of the environment. In comparison with these microscopic elements we are like Swift's giant among the Lilliputians; and this is far from being a sufficiently forcible comparison.

Kinetic Conception of Molecular Motion.—The idea of this peculiar form of motion is by no means new to us. We were familiarized with it in scientific theories during our school days. The atomic theory teaches us that matter behaves, from a chemical point of view, as if it were divided into molecules and atoms. The kinetic theory explains the constitution of gases and the effects of heat by supposing that these particles are endowed with movements of rotation and displacement. The wave theory explains photic phenomena by supposing peculiar vibratory movements in a special medium—the ether. But these are merely hypotheses which are not at all necessary; they are the images of things, not the things themselves.

Reality of the Motion of Particles.—Here there is no question of hypotheses. This internal agitation, this interior labour, this incessant activity of matter are positive facts, an objective reality. It is true that when the chemical or mechanical equilibrium of bodies is disturbed it is only restored more or less slowly. Sometimes days and years are required before it is regained. Scarcely do they attain this relative repose when they are again disturbed, for the environment itself is not fixed; it experiences variations which react in their turn upon the body under consideration; and it is only at the end of these variations, at the end of their respective periods, that they will attain together, in a universal uniformity, an eternal repose.

We shall see that metallic alloys undergo continual physical and chemical changes. They are always seeking a more or less elusive equilibrium. Physicists in modern times have given their attention to this internal activity of material bodies, to the pursuit of stability. Wiedemann, Warburg, Tomlinson, MM. Duguet, Brillouin, Dubem, and Bouasse have revived the old experimental researches of Coulomb and Wertheim on the elasticity of bodies, the effects of pressures and thrusts, the hammering, tempering, and annealing of metals.

The internal activity manifested under these circumstances presents quite remarkable characteristics which cannot but be compared to the analogous phenomena presented by living bodies. Thus have arisen even in physics, a figurative terminology, and metaphorical expressions borrowed from biology.

Comparison of the Activity of Particles with Vital Activity.—Since Lord Kelvin first spoke of the *fatigue* of metals, or the *fatigue* of elasticity, Bose has shown in these same bodies the fatigue of electrical contact. The term *accommodation* has been employed in the study of torsion, and according to Tomlinson for the very phenomena which are the inverse of those of fatigue. The phenomena presented by glass when acted on by an external force which slowly bends it, have been called facts of adaptation. The manner in which a bar of steel resists wire-drawing has been compared to *defensive* processes against threatened rupture. And M. C. E. Guillaume speaks somewhere of "the heroic resistance of the bar of nickel-steel." The term "defence" has also been applied to the behaviour of chloride or iodide of silver when exposed to light.

There has been no hesitation in using the term "memory" concurrently with that of hysteresis to designate the behaviour of bodies acted on by magnetism or by certain mechanical forces. It is true that M. H. Bouasse protests in the name of the physico-mathematicians against the employment of these figurative expressions. But has he not himself written "a twisted wire is a wound-up watch," and elsewhere, "the properties of bodies depend at every moment upon all anterior modifications"? Does not this imply that they retain in some manner the impression of their past evolution? Powerful deformative agencies leave a trace of their action; they modify the body's condition of molecular aggregation, and some physicists go so far as to say that they even modify its chemical constitution. With the exception of M. Duhem, the disciples of the mechanical school who have studied elasticity admit that the effect of an external force upon a body depends upon the forces which have been previously acting on it, and not merely upon those which are acting on it at the present moment. Its present state cannot be anticipated, it is the recapitulation of preceding states. The effect of a torsional force upon a new wire will be different from that of the same force upon a wire previously subjected to torsions and detorsions. It was with reference to actions of this kind that Boltzmann, in 1876, declared that "a wire that has been twisted or drawn out remembers for a certain time the deformations which it has undergone." This memory is obliterated and disappears after a certain definite period. Here then, in a problem of static equilibrium, we find introduced an unexpected factor—time.

To sum up, it is the physicists themselves who have indicated the correspondence between the condition of existence in many brute bodies and that in many living bodies. It cannot be expected that these analogies will in any way serve as explanations. We should rather seek to derive the vital from the physical phenomenon. This is the sole ambition of the physiologist. To derive the physical from the vital phenomenon would be unreasonable. We do not attempt to do this here. It is nevertheless true that analogies are of service, were it only to shake the support which, from the time of Aristotle, has been accorded to the division of the bodies of nature into *psuchia* and *apsuchia*—*i.e.*, into living and brute bodies.

§ 2. The Brownian Movement.

The Existence of the Brownian Movement.—The simplest way of judging of the working activity of matter is to observe it when the liberty of the particles is not interfered with by the action of the neighbouring particles. We approximate to this condition when we watch, through the microscope, grains of dust suspended in a liquid, or globules of oil suspended in water. Now what we see is well known to all microscopists. If the granulations are sufficiently small, they seem to be never at rest. They are animated by a kind of incessant tremor; we see the phenomena called the "Brownian movement." This movement has struck all observers since the invention of the magnifying glass or simple microscope. But the English botanist, Brown, in 1827, made it the object of special research and gave it his name. The exact explanation of it remained for a long time obscure. It was given in 1894 by M. Gouy, the learned physicist of the Faculty of Lyons.

The observer who for the first time looks through the microscope at a drop of water from the river, from the sea, or from any ordinary source—that is to say, water not specially purified—is struck with surprise and admiration at the motion revealed to him. Infusoria, microscopic articulata, and various micro-organisms people the microscopic field, and animate it by their movements; but at the same time all sorts of particles are also agitated, particles which cannot be considered as living beings, and which are, in fact, nothing but organic detritus, mineral dust, and debris of every description. Very often the singular movements of these granulations, which simulate up to a certain point those of living beings, have perplexed the observer or led him to erroneous conclusions, and the bodies have been taken for animalcules or for bacteria.

Characters of this Movement.—But it is as a rule quite easy to avoid this confusion. The Brownian movement is a kind of oscillation, a stationary, dancing to-and-fro movement. It is a Saint Vitus's dance on one and the same spot, and is thus distinguished from the movements of displacement customary with animate beings. Each particle has its own special dance. Each one acts on its own account, independently of its neighbour. There is, however, in the execution of these individual oscillations a kind of common and regular character which arises from the fact that their amplitudes differ little from each other. The largest particles are the slowest; when above four thousandths of a millimetre in diameter, they almost cease to be mobile. The smallest are the most active. When so small as to be barely visible in the microscope, the movement is extremely rapid, and can only occasionally be perceived. It is probable that it would be still more accelerated in smaller objects; but the latter will always escape our observation.

Its Independence of the Nature of the Bodies and of the Environment.—M. Gouy remarked that the movement depends neither on the nature nor on the form of the particles. Even the nature of the liquid has but little effect. Its degree of viscosity alone comes into play. The movements are, indeed, more lively in alcohol or ether, which are very mobile liquids; they are slow in sulphuric acid and in glycerine. In water, a grain one two-thousandth of a millimetre in diameter traverses, in a second, ten or twelve times its own length.

The fact that the Brownian movement is seen in liquors which have been boiled, in acids and in concentrated alkalies, in toxic solutions of all degrees of temperature, shows conclusively that the phenomenon has no vital significance; that it is in no way connected with vital activity so called.

Its Indefinite Duration.—The most remarkable character of this phenomenon is its permanence, its indefinite duration. The movement never ceases, the particles never attain repose and equilibrium. Granitic rocks contain quartz crystals which, at the moment of their formation, include within a closed cavity a drop of water containing a bubble of gas. These bubbles, contemporary with the Plutonian age of the globe, have never since their formation ceased to manifest the Brownian movement.

Its Independence of External Conditions.—What is the cause of this eternal oscillation? Is it a tremor of the earth? No! M. Gouy saw the Brownian movement far away from cities, where the mercurial mirror of a seismoscope showed no subterranean vibration. It does not increase when the vibrations occur and become quite appreciable. Neither is it changed by variation in light, magnetism, or electric influences; in a word, by any external occurrences. The result of observation is to place before us the paradox of a phenomenon which is kept up and indefinitely perpetuated in the interior of a body without known external cause.

The Brownian Movement must be the First Stage of Molecular Motion.—When we take in our hands a sheet of quartz containing a gaseous inclusion, we seem to be holding a perfectly inert object. When we have placed it upon the stage of the microscope, and have seen the agitation of the bubble, we are convinced that this seeming inertia is merely an illusion.

Repose exists only because of our limited vision. We see the objects as we see from afar a crowd of people. We perceive them only as a whole, without being able to discern the individuals or their movements. A visible object is, in the same way, a mass of particles. It is a molecular crowd. It gives us the impression of an indivisible mass, of a block in repose.

But as soon as the lens brings us near to this crowd, as soon as the microscope enlarges for us the minute elements of the brute body, then they appear to us, and we perceive the continual agitation of those elements which are less than four thousandths of a millimetre in diameter. The smaller the particles under consideration, the more lively are their movements. From this we infer that if we could perceive molecules, whose probable dimensions are about one thousand times less, their probable velocity would be, as required by the kinetic theory, some hundreds of metres per second. In the case of objects we can only just see, the Brownian velocity is only a few thousandths of a millimetre per second. No doubt, concludes M. Gouy, the particles that show this velocity are really enormous when compared with true molecules. From this point of view the Brownian movement is but the first degree, and a magnified picture of the molecular vibrations assumed in the kinetic theory.

§ 3. THE INTERNAL ACTIVITY OF BODIES.

Migration of Material Particles.—In the Brownian movement we take into account only very small, isolated masses, small free fragments—*i.e.*, material particles which are not hampered by their relations to neighbouring particles. Any one but a physicist might believe that in true solids endowed with cohesion and tenacity, in which the molecules were bound one to the other, in which form and volume are fixed, there could be no longer movements or changes. This is a mistake. Physics teaches us the contrary, and, in late years especially, has furnished us characteristic examples. There are real migrations of material particles throughout solid bodies—migrations of considerable extent. They are accomplished through the agency of diverse forces acting externally—pressures, thrusts, torsions; sometimes under the action of light, sometimes under the action of electricity, sometimes under the influence of forces of diffusion. The microscopic observation of alloys by H. and A. Lechatelier, J. Hopkinson, Osmond, Charpy, J. R. Benoit; researches into their physical and chemical properties by Calvert, Matthiessen, Riche, Roberts Austen, Lodge, Laurie, and C. E. Guillaume; experiments on the electrolysis of glass, and the curious results of Bose upon electrical contact of metals, show in a striking manner the chemical and kinetic evolutions which occur in the interior of bodies.

Migration under the Action of Weight.—An experiment by Obermeyer, dating from 1877, furnishes a good example of the motions of solid bodies through a hardened viscid mass, taking place under the influence of weight. The black wax that shoemakers and boatbuilders use, is a kind of resin extracted from the pine and other resinous trees, melted in water, and separated from the more fluid part which rises from it. Its colour is due to the lampblack produced by the combustion of straw and fragments of bark. At an ordinary temperature it is a mass so hard that it cannot always be easily scratched by the finger-nail; but if it is left to itself in a receptacle, it finally yields, spreads out as if it were a liquid, and conforms to the shape of the vessel. Suppose we place within a cavity hollowed out of a piece of wood a portion of this substance, and keep it there by means of a few pebbles, having previously placed at the bottom of the cavity a few fragments of some light substance, such as cork. The piece of wax is thus between a light body below and a heavy body above. If we wait a few days, this order is reversed—the wax has filled the cavity by conforming to it; the cork has passed through the wax and appears on the surface, while the stones are at the bottom. We have here the celebrated experiment of the flask with the three elements, in which are seen the liquids mercury, oil, and water superposed in the order of their density, but in this case demonstrated with solid bodies.

Influence of Diffusion.—Diffusion, which disseminates liquids throughout each other, may also cause solids to pass through other solids. Of this W. Roberts Austen gave a convincing proof. This ingenious physicist placed a little cylinder of lead upon a disc of gold, and kept the whole at the temperature of boiling water. At this temperature both metals are perfectly solid, for the melting point of gold is 1,200° C., and of lead is 330°. Still, after this contact has been prolonged for a month and a half, analysis shows that the gold has become diffused through the top of the cylinder of lead.

Influence of Electrolysis.—Electrolysis offers another no less remarkable means of transportation. By its means we may force metals, such as sodium or lithium, through glass walls. The experiment may be performed as indicated by M. Charles Guillaume. A glass bulb containing mercury is placed in a bath of sodium amalgam, and a current is then made to pass from within outward. After some time it can be shown that the metal has penetrated the wall of the bulb, and has become dissolved within it.

Influence of Mechanical Pressure.—Mechanical pressure is also capable of causing one metal to pass into another. We need not recall the well-known experiment of Cailletet, who, by employing considerable pressure, caused mercury to sweat through a block of iron. In a more simple manner W. Spring showed that a disc of copper could be welded to a disc of tin by pressing them strongly one against the other. Up to a certain distance from the surfaces of contact a real alloy is formed; a layer of bronze of a certain thickness unites the two metals, and this could not take place did not the particles of both metals mutually interpenetrate.

§ 4. INTERNAL ACTIVITY OF ALLOYS.

Structure of Alloys.—Metallic alloys have a remarkable structure, which is essentially mobile, and which we have only now begun to understand by the aid of the microscope. Microscopical examination justifies to a certain degree Coulomb's conjecture. That illustrious physicist explained the physical properties of metals by imagining them to be formed of two kinds of elements—integral particles, to which the metal owes its elastic properties, and a *cement* which binds the particles, and to which it owes its coherence. M. Brillouin has also taken up this hypothesis of duality of structure. The metal is supposed to be formed of very small, isolated, crystalline grains, embedded in an almost continuous network of viscous matter. A more or less compact mass surrounding more or less distinct crystals is the conception which may be formed of an alloy.

Changes of Structure produced by Deforming Agencies.—It has been shown that profound changes of crystalline structure can be produced by various mechanical means, such as hammering, and the stretching of metallic bars carried to the point of rupture. Some of these changes are very slow, and it is only after months and years that they are completed, and the metal attains the definite equilibrium corresponding to the conditions to which it is exposed. Though there may be discussions concerning the extent of the transformations to which it is subjected, though some believe they affect the chemical condition of the alloy, while others limit its power to physical effects, it is nevertheless true—and this brings us back to our subject—that the mass of these metals is at work, and that it only slowly attains the phase of complete repose.

The Slow Re-establishment of Equilibrium. Residual Effect.—These operations by which the physical characters of metals are changed, and by which they are adapted to a variety of industrial needs—compression, hammering, rolling, stretching, and torsion—have an immediate, very apparent effect; but they have also a consecutive effect, slowly produced, much less marked and less evident. This is the "residual effect," or "Nachwirkung" of the Germans. It is not without importance, even in practical applications.

Heat also creates a kind of *forced equilibrium.* This becomes but slowly modified, so that a body may remain for a long time in a state which is, however, not the most stable for the conditions under which it is considered. The number of these bodies *not in equilibrium* is as great as that of the substances which have been exposed to fusion. All the Plutonic rocks are in this condition. Glass presents a condition of the same kind. Thermometers placed in melting ice do not always mark the zero Centigrade. This displacement of the zero point falsifies all records if care is not taken to correct it. The correction usually requires prolonged observation. The theory of the displacement of the thermometric zero is not entirely established; but we may suppose, with the author of the *Traité de Thermométrie*, that in glass, as in alloys, are to be found compounds which vary according to the temperature. At each temperature glass tends to assume a determinate composition and a corresponding state of equilibrium; but the previous temperature to which it has been subjected clearly has an influence on the rapidity with which it attains its state of repose. The effect of variation is more marked when we observe glass of more complicated composition. We can understand that those which contain comparable quantities of the two alkalies, soda and potash, may be more subject to these modifications than those having a more simple composition based on a single alkali.

Effects of Annealing.—A piece of brass wire that has been drawn and then heated is the scene of certain very remarkable internal changes, and these have been only recently recognized. The violent treatment of the metallic thread in forcing it through the hole in the die has crushed the crystalline particles; the interior state of the wire is that of broken crystals embedded in a granular mass. Heating changes all that. The crystals separate, repair themselves, and are built up again; they are then hard, geometrical bodies, in an amorphous, relatively soft and plastic mass; their number keeps on increasing; equilibrium is not established until the entire mass is crystallized. We may imagine how many displacements, enormous when compared with their dimensions, the molecules have to undergo when passing through the resisting mass, and arranging themselves in definite places in the crystalline structures.

In the same way, too, in the manufacture of steel, the particles of coal at first applied to the surface pass through the iron.

This *faculty of molecular displacement* enables the metal in some cases to modify its state at one point or another. The use made of this faculty under certain circumstances is very curious, greatly resembling the adaptation of an animal to its environment, or the methods of defence against agents that might destroy it.

Effect of Stretching. Hartmann's Experiment.—When a cylindrical rod of metal, held firmly at either end—a test-piece, as it is called in metallurgy—is pulled sufficiently hard, it often elongates considerably, part of the elongation disappearing as soon as the strain ceases, and the other part remaining. The total elongation is thus the sum of an "elastic elongation," which is temporary, and a "permanent elongation." If we continue the stretching, there appears at some point of the rod a local extension with contraction of sectional area. It is here that the rod will break.

But in place of continuing the stretching, Mr. Hartmann suspends it. He stops, as if to give the "metal-being" time to rally. During this delay it would seem that the molecules hasten to the menaced point to reinforce and harden the weak part. In fact the metal, which was soft at other points, at this spot looks like tempered metal. It is no longer extensible.

When the experimenter begins the stretching again after this rest, and after the narrowed bar has been rolled and become cylindrical again, the local extension and sectional contraction is forced to occur at another point. If another rest is given at this point the metal will also become hardened.

If we repeat the experiment a sufficient number of times, we shall find a total transformation of the rod, which becomes hardened throughout its entire extent. It will break rather than elongate if the stretching is sufficiently severe.

Nickel Steels—their "Heroic" Resistance.—Nickel steels present this phenomena in an exaggerated degree. The alternation of operations which we have just described, bringing the various parts of an ordinary steel rod into a tempered state, is not necessary with nickel steel. The effect is produced in the course of a single trial. As soon as there is any tendency to contraction the alloy hardens at that precise place; the contraction is hardly noticeable; the movement is stopped at this point to attack another weak point, stops there again and attacks a third, and so on; and, finally, the paradoxical fact appears that a rod of metal which was in a soft state and could be considerably elongated has now become throughout its whole extent as hard, brittle, and inextensible as tempered steel. It is in connection with this point that M. C. E. Guillaume spoke of "heroic resistance to rupture." It would seem, in fact, as if the ferro-nickel bar had reinforced each weak point as it was threatened. It is only at the end of these efforts that the inevitable catastrophe occurs.

Effect of Temperature.—When the temperature changes, it is seen that these ferro-nickel bars elongate or retract, modifying at the same time their chemical constitution. But these effects, like those which occur in the glass bulb of a thermometer, do not occur at once. They are produced rapidly for one part, and more slowly for a small remaining portion. Bars of ferro-nickel which have been kept at the same temperature change gradually in length in the course of a year. Can we find a better proof of internal activity occurring in a substance differing so greatly from living matter?

Nature of the Activity of Particles.—These are examples of the internal activity that occurs in brute bodies. Besides, these facts that we are quoting merely to refute Bichat's assertion relative to the immutability of brute bodies, and to show us their activity, also afford us another proof. They show that this activity, like that of animals, wards off foreign intervention, and that this parrying of the attack, again like that of animals, is adapted for the defence and preservation of the brute mass. So that if we consider of special importance the adaptative, teleological characteristic of vital phenomena, a characteristic which is so easily made too much of in biological interpretations, we may also find it again in the inanimate world. To this end we may add to the preceding examples one more which is no less remarkable. This is the famous case of Becquerel's process for colour-photography.

Colour-Photography.—A greyish plate, treated with chloride or iodide of silver and exposed to a red light, rapidly becomes red. It is then exposed to green light, and after passing through dull and obscure tints it becomes green. To explain this remarkable phenomenon, we cannot improve on the following statement:—The silver salt protects itself against the light that threatens its existence; that light causes it to pass through all kinds of stages of coloration before reducing it; the salt stops at the stage which protects it best. It stops at red, if it is red light that assails it, because in becoming red by reflection it best repels that light—*i.e.*, it absorbs it the least.

It may then be advantageous, for the comprehension of natural phenomena, to regard the transformation of inanimate matter as manifestations of a kind of internal life.

Conclusion. Relations of the Surrounding Medium to the Living Being and the Brute Body.—Brute bodies, then, are not immutable any more than are living bodies. Both depend on the medium that surrounds them, and they depend upon it in the same way. Life brings together, brings into conflict, an appropriate organism and a suitable environment. Auguste Comte and Claude Bernard have taught us that vital phenomena result from the reciprocal action of these two factors which are in close correlation. It is also from the reciprocal action of the environment and the brute body that inevitably result the phenomena which that body presents. The living body is sometimes more sensitive to variations of the ambient medium than is the brute body, but at other times the reverse is the case. For example, there is no living organism as impressionable to any kind of stimulus whatever as the bolometer is to the slightest variations of temperature.

There can only be, then, one chemically immutable body—namely, the atom of a simple body, since, by its very definition, it remains unaltered and intangible in combinations. This notion of an unalterable atom has, however, itself been attacked by the doctrine of the ionization of particles due to Sir J. J. Thomson;

and besides, with very few exceptions—those of cadmium, mercury, and the gases of the argon series—the atoms of simple bodies cannot exist in a free state.

Thus, as in the vital struggle, the ambient medium by means of alimentation furnishes to the living being, whether whole or fragmentary, the materials of its organization and the energies which it brings into play. It also furnishes to brute bodies their materials and their energies.

It is also said that the ambient medium furnishes to the living being a third class of things, the *stimuli* of its activities—*i.e.*, its "provocation to action." The protozoon finds in the aquatic environment which is its habitat the stimuli which provoke it to move and to absorb its food. The cells of the metazoon encounter in the same way in the lymph, the blood, and the interstitial liquids which bathe them, the shock, the stimulus which brings their energies into play. They do not derive from themselves, by a mysterious spontaneity without parallel in the rest of nature, the capricious principle which sets them in motion.

Vital spontaneity, so readily accepted by persons ignorant of biology, is disproved by the whole history of the science. Every vital manifestation is a response to a stimulus, a provoked phenomenon. It is unnecessary to say this is also the case with brute bodies, since that is precisely the foundation of the great principle of the inertia of matter. It is plain that it is also as applicable to living as to inanimate matter.

CHAPTER V.

SPECIFIC FORM. LIVING BODIES AND CRYSTALS.

§ 1. Specific form and chemical constitution—The wide distribution of crystalline forms—Organization of crystals—Law of relation between specific form and chemical constitution—Value of form as a characteristic of brute and living beings—Parentage, living beings and mineral parentage—Iso-morphism and the faculty of cross-breeding—Other analogies. § 2. Acquisition and re-establishment of the specific form—Mutilation and regeneration of crystals—Mechanism of reparation.

§ 1. *Specific Form and Chemical Constitution.*—In the enumeration which we have made of the essential features of vitality there are three that are, so to speak, of the highest value. They are, in the order of their importance:—The possession of a specific form; the faculty of growth or nutrition; and finally, the faculty of reproduction by generation. By restricting our comparison between brute bodies and living bodies to these truly fundamental characters we sensibly restrict the field, but we shall see that it does not disappear.

Wide Distribution of Crystalline Forms.—The consideration of specific forms shows us that in the mineral world we need only consider crystallized bodies, as they are almost the only ones that possess definite form. In restricting ourselves to this category we do not limit our field as much as might be supposed. Crystalline forms are very widely distributed. They are, in a measure, universal. Matter has a decided tendency to assume these forms whenever the physical forces which it obeys act with order and regularity, and when their action is undisturbed by accidental occurrences. In the same way, too, living forms are only possible in regulated environments, under normal conditions, protected from cataclysms and convulsions of nature.

The possession of a specific form is the most significant feature of an organized being. Its tendency, from the time it begins to develop from the germ, is toward the acquirement of that form. The progressive manner in which it seeks to realize its architectural plan in spite of the obstacles and difficulties that arise—healing its wounds, repairing its mutilations—all this, in the eyes of the philosophical biologist, forms what is perhaps the most striking characteristic of a living being, that which best shows its unity and its individuality. This property of organogenesis seems pre-eminently the vital property. It is not so, however, for crystalline bodies possess it in an almost equal degree.

The parallel between the crystal and a living being has been often drawn. I will not reproduce it here in detail. My sole desire, after sketching its principal features, is to call attention to the new information that has been brought out by recent investigations.

Organization of Crystals. Views of Haüy, Delafosse, Bravais, and of Wallerant.—In botany, zoology, and crystallography we understand by form an assemblage of material constituents co-ordinated in a definite system—*i.e.*, the organization itself. The body of man, for example, is an edifice in which sixty trillion cells ought each to find its own predetermined place.

In crystallography also we understand by form the organization which crystals present. The grouping of the elements of crystals is, perhaps, more simple. They are none the less organized, in the same sense that living bodies are.

Their organization, while more uniform than that of living bodies, still shows a considerable amount of variation. It should not be assumed that the area of a crystal is completely filled, with contiguous parts applied one to the other by plane faces, as might be supposed from the phenomenon of cleavage which dissociates the parts of the crystalline body into solids of this kind. In reality, the constituent parts are separated from each other by spaces. They are arranged in a quincunx, as Haüy put it, or along the lines of a network, to use the terms of Delafosse and Bravais. The intervals left between them are incomparably larger than their diameters. So that in the organization of a crystal it is necessary to take into account two quite different things :—An element, the crystalline particle, which is a certain aggregate of chemical molecules having a determinate geometrical form ; and a more or less regular, parallelopipedic network, along the edges of which are arranged in a constant and definite manner the aforesaid particles. The external form of the crystal indicates the existence of the network. Its optical properties depend upon the action of the particles, as Wallerant has shown : Thus we must distinguish in a crystal between two kinds of geometrical figures—that of the network and that of the particle—and their characters of symmetry may be either concordant or discordant.

The crystalline particle, the element of the crystal, is therefore a certain molecular complex that repeats itself identically and is identically placed at the nodes of the parallelopipedic network. It has been given different names well calculated to produce confusion—the crystallographic molecule of Mallard, the complex particle of other authors. Some have separated this element into subordinate elements (the fundamental particles of Wallerant and of Lapparent).

These very general outlines will suffice to show how complex and adjustable is the organization of the crystalline individual, which in spite of its geometric regularity and its rigidity, may be compared with the still more flexible organization of the living element. The mineral individual is more stable, more labile—*i.e.*, less prone to undergo change than is the living individual. We may say with M. Lapparent that "crystallized matter presents the most perfect and stable orderly arrangement of which the particles of bodies are susceptible."

Law of Relation of Specific Form to Chemical Constitution.—Crystallization is a method of acquiring specific form. The geometrical architecture of the mineral individual is but little less wonderful or characteristic than that of the living individual. Its form is the result of the mutual reactions of its substances and of the medium in which it is produced; it is the condition of material equilibrium corresponding to a given situation. This idea of a specific form belonging to a given substance under given conditions must be borne in mind. We may consider it as a kind of principle of nature, an elementary law, which may serve as a point of departure for the explanation of phenomena. A particular substance under identical conditions of environment, must always assume a certain form.

This close linking of substance and form, admitted as a postulate in physical sciences, has been carried into biology by some philosophical naturalists, by M. Le Dantec, for instance.

Let us imitate them for a moment. Let us cease to seek in the living being for the prototype of the crystal; let us, on the contrary, seek in the crystal the prototype of the living being. If we succeed in this, we shall then have found the physical basis of life.

Let us say, then, with the biologists we have mentioned, that the substance of each living being is peculiar to it; that it is specific, and that its form—that is to say its organization—follows from it. The morphology of any being whatever, of an animal—of a setter, for example—or even of a determinate being—of Peter, of Paul—is the "crystalline form of their living matter." It is the only form of equilibrium that can be assumed under the given conditions by the substance of the setter, of Peter, or of Paul, just as the cube is the crystalline form of sea-salt. In this manner these biologists have supposed that they could carry back the problem of living form to the problem of living substance, and at the same time reduce the biological mystery to the physical mystery. I have shown above (Chap. V. pp. 199-204) how far this idea is legitimate, and how far and with what restrictions it may be welcomed and adopted.

Value of Form as a Characteristic of Living and Brute Beings.—However this may be, we may say, without fear of exaggeration, that the crystalline form characterizes the mineral with no less precision than the anatomical form characterizes the animal and the plant. In both cases, form—regarded as a method of distribution of the parts—indicates the individual and allows us to diagnose it with more or less facility.

Parentage of Living Beings and Mineral Parentage.—Still another analogy has been noted. In animals and plants similarity in form indicates similarity in descent, community of origin, and proximity in any scheme of classification. In the same way identity of crystalline form indicates mineral relationship. Substances chemically analogous show identical, geometrically superposable forms, and are thus arranged in family or generic groups recognizable at a glance.

Isomorphism and the Faculty of Cross-breeding.—And further, the possibility in the case of isomorphous bodies, of their replacing each other in the same crystal during the process of formation and of thus mingling, so to speak, their congenital elements, may be compared with the possibility of inter-breeding with living beings of the same species. Isomorphism is thus a kind of faculty of crossing. And as the impossibility of crossing is the touchstone of taxonomic relationship, testing it, and separating stocks that ought to be separated, so the operation of crystallization is also a means of separating from an accidental mixture of mineral species the pure forms which are blended therein. Crystallization is the touchstone of the specific purity of minerals; it is the great process in chemical purification.

Other Analogies.—The analogies between crystalline and living forms have been pushed still further even to the verge of exaggeration.

The internal and external symmetry of animals

and plants has been compared to that of crystals. Transitions or intergradations have been sought between the rigid and faceted architecture of the latter and the flexible structure and curved surface of the former; the utricular form of flowers of sulphur on the one hand, and the geometrical structure of the shells of radiolarians on the other, have shown an exchange of typical forms between the two systems. An effort has even been made to draw a parallel between six of the principal types of the animal kingdom and the six crystalline systems. If carried as far as this, our thesis becomes puerile. Real analogies will suffice. Among these the curious facts of crystalline renewal come first.

§ 2. Cicatrization in Living Beings and in Crystals.

We know that living beings not only possess a typical architecture which they have themselves constructed, but that they defend it against destructive agencies, and that if need arise they repair it. The living organism cicatrizes its wounds, repairs losses of substance, regenerates more or less perfectly the parts that have been removed; in other terms, when it has been mutilated it tends to reconstruct itself according to the laws of its own morphology. This phenomenon of reconstitution or reintegration, these more or less successful efforts to re-establish its form and its integrity, at first appear to be a characteristic feature of living beings. This is not the case.

Mutilation and Re-integration of Crystals.—Crystals—let us say crystalline individuals—show a similar aptitude for repairing their mutilations. Pasteur, in an early work, discussed these curious facts. Other experimenters, Gernez a little later and Rauber more recently, took up the same subject, but could do no more than extend and confirm his observations. Crystals are formed from a primitive nucleus, as the animal is formed from an egg; their integral particles are disposed according to efficient geometrical laws, so as to produce the typical form by a constructive process that may be compared to the embryogenic process which builds up the body of an animal. Now this operation may be disturbed by accidents in the surrounding medium or by the deliberate intervention of the experimenter. The crystal is then mutilated. Pasteur saw that these mutilations repaired themselves. "When," said he, "a crystal from which a piece has been broken off is replaced in the mother liquor, we see that while it increases in every direction by a deposit of crystalline particles, activity occurs at the place where it was broken off or deformed; and in a few hours this suffices not only to build up the regular amount required for the increase of all parts of the crystal, but to re-establish regularity of form in the mutilated part." In other words, the work of formation of the crystal is carried on much more actively at the point of lesion than it would have been had there been no lesion. The same thing would have occurred with a living being.

Mechanism of Reparation.—Gernez some years later made known the mechanism of this reparation, or, at least, its immediate cause. He showed that on the injured surface the crystal becomes less soluble than on the other facets. This is not, however, an exceptional phenomenon. It is, on the contrary, quite frequently observed that the different faces of a crystal show marked differences in solubility. This is what happens in every case for the mutilated face in comparison with the others; the matter is less soluble there. The consequence of this is clear; the growth must preponderate on that face, since there the mother liquor will become super-saturated before being super-saturated for the others. We may explain this result in another way. Each face of the crystal in contact with the mother liquor is exposed to two antagonistic actions: The matter deposited upon a surface may be taken away and redissolved if, for any reason whatever, such matter becomes more soluble than that of the liquid stratum in contact with it; in the second place, the matter of this liquid stratum may, under contrary conditions, be deposited, and thus increase the body of the crystal. There is, then, for each point of the crystalline facet, a positive operation of deposit which results in a gain, and a negative operation of redissolution which results in a loss. One or the other effect predominates according as the relative solubility is greater or less for the matter of the facet under consideration. On the mutilated surface it is diminished, deposition then prevails.

But this is only the immediate cause of the phenomenon; and if we wish to know why the solubility has diminished on the mutilated surface Ostwald explains it to us by showing that crystallization tends to form a polyhedron in which the surface energy is a relative mimimum.

CHAPTER VI.

NUTRITION IN THE LIVING BEING AND IN THE CRYSTAL.

Assimilation and growth in the crystal.—Methods of growth in the crystal and in the living being; intussusception; apposition.—Secondary and unimportant character of the process of intussusception.

I HAVE already stated (Chap. VI. p. 209) that nutrition may be considered as the most characteristic and essential property of living beings. Such beings are in a state of continual exchange with the surrounding medium. They assimilate and dissimilate. By assimilation the substance of their being increases at the expense of the surrounding alimentary material, which is rendered similar to that of the being itself.

Assimilation and Growth in the Crystal.—There exists in the crystal a property analogous to nutrition, a kind of nutrility, which is the rudiment of this fundamental property of living beings. The development of a crystal starts from a primitive nucleus, the germ of the crystalline individual that we will presently compare to the ovum or embryo of a plant or an animal. Placed in a suitable culture-medium—*i.e.*, in a solution of the substance—this germ develops. It assimilates the matter in solution, incorporates the particles of it, and increases, preserving at the same time its form, reproducing its specific type or a variety of it. Its growth proceeds without interruption. The crystalline individual may attain quite a large size if we know how to nourish it properly—we might say, to fatten it. Very frequently, at a given time, a new particle of the crystal serves in its turn as a primitive nucleus, and becomes the point of departure for a new crystal engrafted upon the first.

Taken from its mother liquor, placed where it cannot be nourished, the crystal, arrested in its growth, falls into a condition of rest not without analogy to that of a seed or of a reviviscent animal. Its evolution is resumed with the return of favourable conditions—the bath of soluble matter.

The crystal is in a relation of continual exchange with the surrounding medium which feeds it. These exchanges are regulated by the state of this medium, or, more exactly, by the state of the liquid stratum which is in immediate contact with the crystals. It loses or it gains in substance if, for example, this layer becomes heated or cooled more rapidly than the crystal. In a general way, it assimilates or dissimilates according as its immediate environment is saturated or diluted. Here, then, we have a kind of mobile equilibrium, comparable, in some measure, to that of the living being.

Methods of Growth of the Crystal and of the Living Being. Intussusception. Apposition.—In truth, there seems to be a complete opposition between the crystal and the living being as regards their manner of nutrition and growth. In the one case the method is intussusception; in the other it is apposition. The crystalline individual is all surface. Its mass is impenetrable to the nutritive materials. Since only the surface is accessible, the incorporation of similar particles is possible only by external juxtaposition, and the edifice increases only because a new layer of stones has been added to those which were there before. On the contrary, the body of an animal is a mass essentially penetrable. The cellular elements that compose it have more or less rounded and flexible forms. Their contact is by no means perfect. They have neither the stiffness nor the precision of adjustment that the crystalline particles have. Liquids and gases can insinuate themselves from without and circulate within the meshes of this loose construction. Assimilation can therefore take place throughout its whole depth, and the edifice increases because each stone is itself increasing.

The Secondary and Commonplace Character of the Process of Intussusception.—The apparent opposition of these two processes is doubtless diminished if we compare the simple mineral individual with the elementary living unit, the crystalline particle with the protoplasmic mass of a cell. Without carrying analysis so far as this, it is yet easy to see that apposition and intussusception are mechanical means that living beings employ at one and the same time and combine according to their necessities. The hard parts of the internal and external skeleton increase both by interposition and superposition, at once. It is by the last method that bones increase in diameter, and the shells of molluscs, the scales of reptiles and fishes, and the testae of many radiate animals are formed. In these organs, as in crystals, life and nutrition occur at the surface.

Apposition and intussusception are then secondary, mechanical arrangements having relation to the physical characters of the body—solidity in the crystal, semi-fluidity in the cellular protoplasm. If we compare the inorganic liquid matter with the semi-fluid organized matter, we recognize that the addition of substance takes place in the same manner in each—*i.e.*, by interposition. If we add a soluble salt to a fluid, the molecules of the salt separate themselves and interpose themselves between those of the fluid. There is, therefore, nothing especially mysterious or particularly vital about the process of intussusception. Applied to fluid protoplasm, it is merely the diffusion that ordinarily occurs in mixed liquids.

CHAPTER VII.

GENERATION IN BRUTE BODIES AND LIVING BODIES. SPONTANEOUS GENERATION.

Protoplasm a substance which continues—Case of the crystal—Characteristics of generation in the living being—Property of growth—Supposed to be confined to the living being—Fertilization of micro-organisms—Fertilization of crystals—Sterilization of crystalline and living media—Spontaneous generation of crystals—Metastable and labile zones—Glycerine crystals—Possible extinction of a crystalline species—Conclusion.

WE have not yet exhausted the analogies between a crystal and the living being. The possession of a specific form, the tendency to re-establish it by re-disintegration and the existence of a kind of nutrition are not sufficient to constitute complete similarity. It still lacks a fundamental character, that of generation. Chauffard some time ago, in an attack which he made upon the physiological ideas of his day, aptly exhibited this weak point. "Let us disregard," he said, "those interesting facts relative to the acquisition of a typical form—facts that are common to the mineral world as well as to living beings. It is none the less true that the crystalline type is in no way derived from other pre-existing types, and that nothing in crystallization recalls the actions of ascendants and the laws of heredity."

This gap has since been filled. The work of

Gernez, of Violette, of Lecoq de Boisbaudran, the experiments of Ostwald and of Tammann, the observations of Crookes and of Armstrong—all this series of researches, so lucidly summarized by M. Leo Errera in his essays in botanical philosophy, had for their result the establishment of an unsuspected relation between the processes of crystallization and those of generation in animals and plants.

Protoplasm is a Substance which Continues. The Case of the Crystal.—Under present conditions a living being of any kind springs from another living being similar to itself.

Its protoplasm is always a continuation of the protoplasm of an ancestor. It is an atavic substance of which we do not see the beginning; we only see it continue. The anatomical element comes from a preceding anatomical element, and the higher animal itself comes from a pre-existing cell of the material organism, the ovum. The ladder of filiation reaches back indefinitely into the past.

We shall see that there is something analogous to this in certain crystals. They are born of a preceding individual; they may be considered as the posterity of the antecedent crystal. If we speak of the matter of a crystal as the matter of a living being is spoken of, in cases of this kind we would say that the crystalline substance is an atavic substance of which we see only the continuation, as in the case of protoplasm.

Characters of Generation in the Living Being.—Growth of the living substance, and consequently of the being itself, is the fundamental law of vitality. Generation is the necessary consequence of growth (p. 210).

Living elements or cells cannot subsist indefinitely without increasing and multiplying. The time must come when the cell divides, either directly or indirectly; and then, instead of one cell, there are two. This is the method of generation for the anatomical element. In a complex individual it is a more or less restricted part of the organism, usually a simple sexual cell, that takes on the formation of the new being, and assures the perpetuity of the protoplasm, and therefore of the species.

Property of Growth. Its Supposed Restriction to Living Beings.—At first it would appear that nothing like this occurs in inanimate nature. The physical machine, if we furnish it matter and energy, could go on working indefinitely, without being compelled to increase and reproduce. Here, then, there is an entirely new condition peculiar to the organized being, a property well adapted, it would seem—and this time without any possible doubt—for separating living matter from brute matter. It is not so.

It would not be impossible to imagine a system of chemical bodies organized like the animal or vegetable economy, so that a destruction would be compensated for by a growth. The only thing impossible is to suppose, with M. le Dantec, a destruction that would at the same time be an analysis. And an additional perplexity occurs when he supposes that in the successive acts exchanges of material may occur.

There is no necessity for making this impossible chemistry a characteristic of the living being. The chemistry of the living being is general chemistry. Lavoisier and Berthelot enforced this view. We should not lose sight of the teachings of the masters.

Let us return to generation, properly so called, and find in it the characteristics of brute bodies and of crystals.

The Sowing of Micro-organisms.—When a micro-biologist wishes to propagate a species of micro-organisms, he places in a culture medium a few individuals (one is all that is actually necessary), and soon observes their rapid multiplication. Usually, if only the ordinary microbes in atmospheric dust are wanted, the operator need not trouble to charge the culture; if the culture tube remains open and the medium is suitably chosen, some germ of a common species will fall in and the liquid will become colonized. This has the appearance of spontaneous generation.

The Sowing of Crystals.—Concentrated solutions of various substances, supersaturated solutions of sodium magnesium sulphate, and sodium chlorate are also wonderful culture media for certain mineral organic units—certain crystalline germs. Ch. Dufour, experimenting with water cooled below 0° C., its point of solidification; Ostwald, with salol kept below 39°.5, its point of fusion; Tammann, with betol, which melts at 96°; and, before them, Gernez, with melted phosphorus and sulphur—all these physicists have shown that liquids in superfusion are also media specially appropriate for the culture and propagation of certain kinds of crystalline individuals.

Some of these facts have become classic. Lowitz showed in 1785 that a solution of sodium sulphate could be concentrated by evaporation so as to contain more salt than was conformable with the temperature, without, however, depositing the excess. But if a solid fragment, a crystal of salt, is thrown into the liquor, the whole of the excess immediately passes into the state of a crystallized mass. The first crystal has engendered a second similar to itself; the latter has engendered a third, and so on from one to the other. If we compare this phenomenon with that of the rapid multiplication of a species of microbes in a suitable culture medium, no difference will be perceived. Or perhaps we may note one unimportant difference—the rapidity of the propagation of the crystalline germs as opposed to the relative slowness of the generation of the micro-organisms.

Again, the propagation of crystallization in a supersaturated or superfused liquid may be delayed by appropriate devices. The crystalline individual gives birth, then, to another individual that conforms to its own type, or even to varieties of that type when such exist. Into the right branch of a U tube filled with sulphur in a state of superfusion Gernez dropped octahedric crystals of sulphur, and into the left branch prismatic crystals. On either side were produced new crystals conforming to the type that had been sown.

Sterilization of Crystalline Media and Living Media.—Ostwald varied these experiments by using salol. He melted the substance by heating it above 39°.5 C.; then, protecting it from crystals of any kind, he let the solution stand in a closed tube. The salol remained liquid indefinitely—until it was touched with a platinum wire that had been in contact with solid salol—*i.e.*, until a crystalline germ was introduced. But if the platinum wire has been previously sterilized by passing it, as the bacteriologists do, through a flame, it can then be introduced into the liquor with impunity.

The Dimensions of Crystalline Germs Comparable to those of Microbes.—We may dilute the solid salol with inert powder—lactin, for example—dilute the first mixture with a second, the second with a third, and so on; then, throwing into the solution of surfused salol a tenth of a milligram from one of these various mixtures, we find that the production of crystals will not take place if the fragment thrown in weighs less than a millionth of a milligram, or measures less than ten thousandths of a millimetre in length. It would seem, then, that these are the dimensions of the crystalline particle or crystallographic molecule of salol. In the same way Ostwald satisfied himself that the crystalline germ of hyposulphite of soda weighs about a thousand-millionth of a milligram, and measures a thousandth of a millimetre; that of chlorate of soda weighs a ten-millionth of a milligram. These dimensions are entirely comparable with those of microbes.

All these phenomena have been studied with a detail into which it is impossible to enter here, and which clearly shows more and more intimate analogies between the formation of crystals and the generation of micro-organisms.

Extension and Propagation of Crystallization. Optimum Temperature of Incubation.—Crystallization which has commenced around a germ is propagated more or less rapidly, and ends by invading the whole of the liquor.

The rapidity of this movement of extension depends upon the conditions of the medium, especially upon its temperature. This is shown very well by Tammann's experiments with betol. This body, the salicylic ester of naphthol, fuses at 96° C. If it is melted in tubes sealed at a temperature of 100° C., it may be cooled to lower and lower temperatures—to + 70°, to + 25°, to + 10°, to — 5° without solidifying. Let us suppose that by some combination of circumstances a few centres of crystallization—that is to say, of crystalline germs—have appeared in the solution. Solidification will extend slowly at the ordinary temperature, at 20° to 25° and thereabouts. On the other hand, it will be propagated with great rapidity if the liquor is kept at about 70°. This point—70°—is the thermal optimum for the propagation of germs. It is the most favourable temperature for what may be called their incubation. As soon as the germs find themselves in a liquor at 70° they increase, multiply, and show that they are in the best conditions for growth.

Spontaneous Generation of Crystals. Optimum Temperature for the Appearance of Germs.—If we consider various supersaturated solutions or liquids in superfusion, we shall soon discover that they can be arranged in two categories. Some remain indefinitely liquid under given conditions unless a crystalline germ is introduced into them. Others solidify spontaneously without artificial intervention, and such crystallization may even be propagated very rapidly under determinate conditions. This implies that these are conditions favouring the spontaneous appearance of germs.

This distinction between substances of crystalline generation by filiation and substances of spontaneous crystalline generation is not specific. The same substance may present the two methods of generation according to the conditions in which it is placed. Betol furnishes a good example of this. Liquefy it at

100° in a sealed tube and keep it by means of a stove above 30°, and it will remain liquid almost indefinitely. On the other hand, lower its temperature and leave it for one or two minutes at 10°, and germs will appear in the liquor; prolong the exposure to this degree of heat and the number of these spontaneously appearing germs, appearing in isolation, will rapidly increase. On the other hand, you will observe that propagation by filiation—that is to say, by extension from one to another—is almost absent. The temperature of 10° is not favourable to that method of generation; and we have just seen, in fact, that it is at a temperature of about 70° that extension of crystallization from one to another is best accomplished. The temperature of 70° was the optimum for propagation by filiation. Inversely, the temperature of 10° is the optimum for spontaneous generation. Above and below this optimum the action is slower. We may count the centres of crystallization, which slowly extend further and further, as in a microbic culture one counts the colonies corresponding to the germs primitively formed. To sum up, if there is an optimum for the formation of crystals, there is a different optimum for their rapid extension.

The Metastable and Labile Zones.—This phenomena is general. There is for each substance a set of conditions (temperature, degree of concentration, volume of the solution) in which the crystalline individuals can be produced only by germs or by filiation. This is what occurs for betol above the temperature of 30°. The body is then in what Ostwald has called a *metastable* zone. There is, however, for the same body another set of circumstances more or less complete, in which its gems appear simultaneously. This is what happens for betol at about the temperatute of 10°. These circumstances are those of the *labile zone* or zone of spontaneous generation.

Crystals of Glycerine.—We may go a step further. Let us suppose, with L. Errera, that we have a liquid in a state of metastable equilibrium, whose labile equilibrium is as yet unknown. This is what actually occurs for a very widely known body, glycerine. We do not know under what conditions glycerine crystallizes spontaneously. If we cool it, it becomes viscous; we cannot obtain its crystals in that way. It was not found in crystals until 1867. In that year, in a cask sent from Vienna to London during winter, crystallised glycerine was found, and Crookes showed these crystals to the Chemical Society of London. What circumstances had determined their formation? We knew not then, and we know not now. It may be observed that this case of spontaneous generation of the crystals of glycerine has not remained the solitary instance. M. Henninger has noted the accidental formation of glycerine crystals in a manufactory in St. Denis.

It may be remarked that this crystalline species appeared, as living species may have done, at a given moment in an environment in which a favourable chance combined the necessary conditions for its production. It is also quite comparable to the creation of a living species; for having once appeared we have been able to perpetuate it. The crystalline individuals of 1867 have had a posterity. They have been sown in glycerine in a state of superfusion, and there they reproduced themselves. These generations have been sufficiently numerous to spread the species throughout a great part of Europe. M. Hoogewerf showed a great flask full to the Dutch biologists who met at Utrecht in 1891. M. L. Errera presented others in June 1899, to the Society of Medical and Natural Sciences at Brussels. To-day the great manufactory of Sarg & Co., of Vienna, is engaged in their production on a large scale for industrial purposes.

Thus we are able to study this crystalline species of glycerine and to determine with precision the conditions of its continued existence. It has been shown that it does not resist a temperature of 18°, so that if precautions were not taken to preserve it, a single summer would suffice to annihilate all the crystalline individuals existing on the surface of the globe, and thus the species would be extinguished.

Possible Extinction of a Crystalline Species.—As these crystals melt at 18°, this temperature represents the point of fusion of solid glycerine or the point of solidification of liquid glycerine. But the liquor does not solidify at all if its temperature falls below 18° C., as we well know, for it is at that temperature we use it. Nor does it solidify at zero, nor even at 18° below zero; at 20°, for instance, it merely thickens and becomes pasty. We only know glycerine, then, in a state of superfusion, a fact which chemists have not learned without amazement. Under these conditions, so analogous to the appearance of a living species, to its unlimited propagation and to its extinction, the mineral world offers a quite faithful counterpart to the animal world. The living body illustrates here the history of the brute body and facilitates its exposition. Inversely, the brute body in its turn throws remarkable light on the subject of the living body, and on one of the most serious problems relative to its origin, that of spontaneous generation.

Conclusion.—These facts lead to one conclusion. Until the concourse of propitious circumstances favourable to their spontaneous generation was brought about, crystals were obtained only by filiation. Until the discovery of electro-magnetism, magnets were made only by filiation, by means of the simple or double application of a pre-existing magnet. Before the discovery which fable attributes to Prometheus, every new fire was produced only by means of a spark from a pre-existing fire. We are at the same historical stage as regards the living world, and that is why, up to the present, there has never been formed a single particle of living matter except by filiation, except by the intervention of a pre-existing living organism.

BOOK V.

SENESCENCE AND DEATH.

WE grow old and we die. We see the beings which surround us grow old and disappear. At first we see no exceptions to this inexorable law, and we consider it as a universal and inevitable law of nature. But is this generalization well founded? Is it true that no being can escape the cruel fate of old age and death, to which we and all the representatives of the higher animality are exposed? Or, on the other hand, are any beings immortal? Biology answers that, in fact, some beings are immortal. There are beings to whose life no law assigns a limit, and they are the simplest, the least differentiated and the least perfect. Death thus appears to be a singular privilege attached to organic superiority, the ransom paid for a masterly complexity. Above these elementary, monocellular, undifferentiated beings, which are protected from mortality, we find others, higher in their organization, which are exposed to it, but with whom death seems but an accident, avoidable in principle if not in fact. The anatomical elements of this higher animal are a case in point. Flourens once tried to persuade us that the threshold of old age might be made to recede considerably, and there are biologists in the present day who give us some glimpse of a kind of vague immortality. We may, therefore, ask our readers to follow us in our examination of these re-opened if not novel questions, and we shall explain the views of contemporary physiology as to the nature of death, its causes, its mechanisms, and its signs.

CHAPTER I.

VARIOUS WAYS OF REGARDING DEATH.

Different meanings of the word death—Physiological distinction between elementary and general death — Non-scientific opinions—The ordinary point of view—Medical point of view.—The signs of death are prognostic signs.

Different Meanings of the Word Death.—An English philosopher has asserted that the word we translate by "cause" has no less than sixty-four different meanings in Plato and forty-eight in Aristotle. The word "death" has not so many meanings in modern languages, but still it has many. Sometimes it indicates an action which is taking place, the action of dying, and sometimes a state, the state which succeeds the action of dying. The phenomena it connotes are in the eyes of many biologists quite different, according as we watch them in an animal of complex organization, or on the other hand, in monocellular beings, protozoa and protophytes.

Physiological Distinction between Elementary Death and General Death.—We distinguish the death of the anatomical elements, *elementary death*, from the death of the individual regarded as a whole, *general death.* Hence we recognize an *apparent death*, which is an incomplete and temporary suspension of the phenomena of vitality, and a *real death*, which is a final and total arrest of these phenomena. When we consider it in its essential nature (assumed, but not known) we look on it as the *contrary of life*, as did the Encyclopædia, Cuvier, and Bichat; or we regard it with others either as the consequence of life, or simply as the end of life.

Non-scientific Opinions.—What is death to those outside the realm of science? First of all we find the consoling solution given by those who believe death to be the commencement of another life. We next find ourselves involved in a confused medley, an infinite diversity of philosophical doubt and superstition. "A leap into the unknown," says one. "Dreamless and unconscious night," says another. And again, "A sleep which knows no waking." Or, with Horace, "the eternal exile," or with Seneca, annihilation. *Post mortem nihil; ipsaque mors nihil.*

The idea which is constantly supervening in the midst of this conflict of opinion is that of the *breaking up* of the elements, the union of which forms the living being. It has, as we shall see, a real foundation which may perhaps receive the support of science. We shall not find that the best way of defining death is to say that it consists of the "dissolution of the society formed by the anatomical elements, or again, in the dissolution of the consciousness that the individual possesses of himself—*i.e*, of the existence of this society." It is the rupture of the social bond. The old idea of dispersion is a variant of the same notion. But the ancients evidently could not understand, as we do, the nature of these elements which are associated to form the living being, and which are liberated or dispersed by death. We, as biologists, can see microscopical organic unity with a real objective existence. The ancients were thinking of spiritual elements, of principles, of entities. To the Romans, who may be said to have held that there are three souls, death was produced by their separation from the body. The first, the breath, the *spiritus*, mounting towards celestial regions (*astra petit*); the second, the *shade*, remaining on the surface of the earth and wandering around the tombs; the third, the *manes*, descending to the lower regions. The belief of the Hindoos was slightly different. The body returned to the earth, the breath to the winds, the fire of the glance to the sun, and the ethereal soul to the world of the pure. Such were the ideas of mortal dispersion formed by ancient humanity.

Modern science takes a more objective point of view. It asks by what facts, by what observable events death is indicated. Generally speaking, we may say that these facts interrupt an interior state of things which was life and to which they put an end. Thus death is defined by life. It is the cessation of the events and of the phenomena which characterize life. We must, therefore, know what life is to understand the meaning of death. How wise was Confucius when he said to his disciple, Li-Kou:—"If we do not know life, how can we know death?" According to biology there are two kinds of death because there are two kinds of life; elementary life and death correspond just as general life and death do, and this is where scientific opinion diverges from commonly received opinion.

What cares the man who reasons as most human beings do, about this life of the anatomical elements of his body, the existence and the silent activity of which are in no way revealed to him. What does their death matter to him? To him there is but one poignant question, that of being separated or not being separated from the society of his fellows. Death is no longer to feel, no longer to think; it is the assurance that one will never feel, one will never think again. Sleep, dreamless sleep, is already in our eyes a kind of transient death; but, when we fall asleep we are sure of waking again. There is no awaking from the sleep of death. But that is not all. Man knows that death, this dreamless sleep that knows no waking, will be followed by the dissolution of his body. And what a dissolution will there be for the body, the object of his continual care! Remember the description of Cuvier—the flesh that passes from green to blue and from blue to black, the part which flows away in putrid venom, the other part which evaporates in foul emanations, and finally, the few ashes that remain, the tiny pinch of minerals, saline or earthy, which are all that is left of that once animated masterpiece.

The Popular View.—To the man afraid of death it seems, in the presence of so great a catastrophe, that the patient analysis of the physiologist scrupulously noting the succession of phenomena and explaining their sequence is uninteresting. He will only attach the slightest importance to knowing that vestiges of vitality remain in this or that part of his body, if they do not re-establish in every part the *status quo ante*. He cares not to hear that a certain time after the formal declaration of his death his nails and his hair will continue to grow, that his muscles will still have the useless faculty of contraction, that every organ, every tissue, every element, will oppose a more or less prolonged resistance to the invasion of death.

Medical View.—It is, however, these very facts and details, this why and wherefore, which interest the physiologist. The state of mind of the doctor in this respect, again, is different. When, for instance, the doctor declares that such and such a person is dead, he is really making not so much a statement of fact as a prediction. How many elements are still living and will be capable of new birth in this corpse that he has before his eyes? That is not what he asks himself, nor is it what we should ask of him. He knows, besides, that all these partial survivals will be extinguished and will never find the conditions necessary to reviviscence, and that the organization will never be restored to its primal activity; and this is what he affirms. The fear of premature burial which haunts so many imaginations is the fear of an error in the prediction. It is to avoid this that practical medicine has devoted so much of its attention to the discovery of a *certain*—and early—sign of death. By this we understand the discovery of a *certain prognostic sign of general death.* We want a prognostic sign enabling us to assert that the life of the brain is now extinguished and will never be reanimated. And yet there are in that organism many elements which are still alive. Many others even may be born anew if we could give them suitable conditions which they no longer meet with in the animal machine now thrown out of gear. What finer example could we give than the experiment of Kuliabko, the Russian physiologist, who kept a man's heart working and beating for eighteen hours after the official verification of his death.

21

CHAPTER II.

THE PROCESS OF DEATH.

Constitution of organisms.—Partial lives.—Collective life.—The rôle of apparatus.—Death by lesion of the major apparatus.—The vital tripod.—Solidarity of the anatomical elements.—Humoral solidarity.—Nervous solidarity.—Independence and subordination of the anatomical elements.

Partial Lives. Collective Life.—With the exception of the physiologist, no one, neither he who is ignorant nor he who is intellectual, nor even the doctor, troubles his head about the life or the death of the element, although this is the basis, the real foundation, of the activity manifested by the social body and by its different organs. The life of the individual, of the animal, depends on these elementary partial lives just as the existence of the State depends upon that of its citizens. To the physiologist, the organism is a federation of cellular elements unified by close association. Goethe compared them to a "multitude"; Kant to a "nation"; and others have likened them to a populous city the anatomical elements of which are the citizens, and which possesses an individuality of its own. So that the activity of the federated organism may be discussed in each of its parts, and then it is *elementary life*, or in its totality, and then it is *general life*. Paracelsus and Bordeu had a glimpse of this truth when they considered a life appropriate to each part (*vita propria*) and a collective life, the life of the whole (*vita communis*). In the same way we must distinguish the *elementary death*, which is the cessation of the vital phenomena in the isolated cell, from the *general death*, which is the disappearance of the phenomena which characterised the collectivity, the totality, the federation, the nation, the city, the whole in so far as it is a unit.

These comparisons enable us to understand how general life depends on the partial lives of each anatomical citizen. If all die, the nation, the federation, the total being clearly ceases to exist. This city has an enormous population—there are thirty trillion cellules in the body of man; it is peopled with absolutely sedentary citizens, each of which has its fixed place, which it never leaves, and in which it lives and dies. It must possess a system of more or less perfect arrangements to secure the material life of each inhabitant. All have analogous requirements: they feed very much the same; they breathe in the same way; each in fact has its profession, industry, talents, and aptitudes by which it contributes to social life, and on which, in its turn, it depends. But the process of alimentation is the same for all. They must have water, nitrogenous materials and analogous ternaries; the same mineral substances, and the same vital gas, oxygen. It is no less necessary that the wastes and the egesta, very much alike in every respect, should be carried off and borne away in discharges arranged so as to free the whole system from the inconvenience, the unhealthiness, and the danger of these residues.

Secondary Organization in Organs.—That is why, as we said above, the secondary organizations of the economy exist:—the digestive apparatus which prepares the food and enables it to pass into the blood, into the lymph, and finally into the liquid medium which bathes each cell and constitutes its real medium; the respiratory apparatus which imports the oxygen and exports the gaseous excrement, carbonic acid; the heart and the circulatory system which distributes through the system the internal medium, suitably purified and recuperated. The organization is dominated by the necessities of cellular life. This is the law of the city, to which Claude Bernard has given the name of the *law of the constitution of organisms.*

Death by Lesion of the Major Organs. Vital Tripod.—Thus we understand what life is, and at the same time what is the death of a complex living being. The city perishes if its more or less complicated mechanisms which look after its revictualling and its discharge are seriously affected at any point. The different groups may survive for a more or less lengthy period, but progressively deprived of the means of food or of discharge, they are finally involved in the general ruin. If the heart stops, there is a universal famine; if the lungs are seriously injured, we are asphyxiated; if the principal organ of discharge, the kidney, ceases to perform its allotted task, there is a general poisoning by the used-up and toxic materials retained in the blood.

We understand how the integrity of the major organs,—the heart, the lungs, the kidney,—is indispensable to the maintenance of existence. We understand that their lesion, by a series of successive repercussions, involves universal death. We always die, said the doctors of old, because of the failure of one of these three organs, the heart, the lungs, or the brain. Life, they said in their inaccurate language, depends upon these as upon three supports. Hence the idea of the *vital tripod.* But it is not only this trio of organs which maintain the organism; the kidney and the liver are no less important. In different degrees each part exercises its action on the rest. Life is based in reality on the immense multitude of living cells associated for the formation of the body; on the thirty trillion anatomical elements, each part is more or less necessary to all the rest, according as the bond of solidarity is drawn more or less closely in the organism under consideration.

Death and the Brain.—There are indeed more noble elements charged with higher functions than the rest. These are the nervous elements. Those of the brain preside over the higher functions of animality, sensibility, voluntary movement, and the exercise of the intellect. The rest of the nervous system forms an instrument of centralization which establishes the relations of the parts one with the other and secures their solidarity. When the brain is stricken and its functions cease, man has lost the consciousness of his existence. Life seems to have disappeared. We say of a man in this plight that he no longer lives, thus confusing general life with the cerebral life which is its highest manifestation. But the man or the animal without a brain lives what may be called a vegetative life. The human anencephalic foetus lives for some time, just as the foetus which is properly formed. Observation always shows that this existence of the other parts of the body cannot be sustained indefinitely in the absence of that of the brain. By a series of impulses due to the solidarity of the grouping of the parts, the injury received by the brain affects by repercussion the other organs, and leads in the long run to the arrest of elementary life in all the anatomical elements. The death of the whole is then complete.

Doctors have therefore a two-fold reason for saying that the brain may cause death. The death of the brain suppresses the highest manifestation of life, and, in the second place, by a more or less remote counter stroke, it suppresses life in all the rest of the system.

Death is a Process.—Besides, the fact is general. The death of one part always involves the death of the rest—*i.e.*, universal death. A living organism cannot be at the same time alive and a cemetery. The corpses cannot exist side by side with the living elements. The dead contaminates the living, or in some other way involves it in its ruin. Death is propagated; it is a progressive phenomenon which begins at one point and gradually is extended to the whole. It has a beginning and a duration. In other words, the death of a complex organism is a process. And further, the end of a simple organism, of a protozoan, of a cell, is itself a process infinitely more shortened.

The very perfection of the organism is therefore the cause of its fragility. It is the degree of solidarity of the parts one with another which involves the one set in the catastrophe of the rest, just as in a delicate piece of mechanism the derangement of a wheel brings nearer and nearer the total breakdown. The important parts, the lungs, the heart, the brain, suffer no serious alteration without the reflex being felt throughout. But there are also wheels less evident, the integrity of which is scarcely less necessary.

The Solidarity of the Anatomical Elements.—The cause of the mortal process—*i.e.*, of the extension and the propagation of an initial destruction—is therefore to be found in the solidarity of the parts of the organism. The closer it is the greater do the chances of destruction become, for the accident which has happened to one will by repercussions affect the others.

Now the solidarity of the parts of the organism may be carried out in two ways; there is a *humoral solidarity* and a *nervous solidarity.*

Humoral Solidarity.—Humoral solidarity is realized by the mixture of humours. All the liquids of the organism which have lodged in the interstices of the elements and which soak the tissues, are in contact and in relation of exchange one with another, and through the permeable wall of the small vessels they are in relation with the blood and the lymph.

All the liquid atmospheres which surround the cells and form their ambient medium have intercommunication. A change having taken place in one cellular group, and therefore in the corresponding liquid, modifies the medium of the further or nearer groups, and therefore these groups themselves.

Nervous Solidarity.—But the real instrument of the solidarity of the part is the nervous system. Thanks to it in the living machine the component activities of the cellular multitude restrain and control one another. Nervous solidarity makes of the complex being not a mob of cells, but a connected system, an individual in which the parts are subordinated to the whole and the whole to the parts; in which the social organism has its rights just as the individual has his rights. The whole secret of the vital functional activity of the complex being is contained in these two factors:—the independence and the subordination of the elementary lives. General life is the harmony of the elementary lives, their symphony.

Independence and Subordination of the Anatomical Elements.—The independence of the anatomical elements results from the fact that they are the real depositaries of the vital properties, the really active components. On the other hand the subordination of the parts to the whole is the very condition of the preservation of form in animals and plants. The architecture which is characteristic of them, the morphological plan which they realize in their evolutive development which they are ever preserving and repairing, form a striking proof of this. This dependence in no way contradicts the autonomy of the elements. For when with Claude Bernard and Virchow we study the circumstances we see that the element accommodates itself to the organic plan without violence to its nature. It behaves in its natural place as it would behave elsewhere, if elsewhere it were to meet around it the same liquid medium which at once is a stimulant and a food. This at least is the conclusion we may draw from experiments on transplanting, or on animal and vegetable grafting. Neither the neighbouring elements, nor the whole system act on it at a distance by a kind of mysterious induction, according to the ideas of the vitalists, in order to regulate the activity of the element. They contribute solely to the composition of the liquid atmosphere which bathes it. They intervene in order to provide it with a certain environment whose very characteristic physical and chemical constitution regulates its activity. This constitution may be some day imitated by the devices of experiment. When that result is achieved the anatomical element will live in isolation exactly as it lives in the organic association, and the mysterious bond which causes its solidarity with the rest of the economy will become intelligible. In fact, we may defer more or less the maturity of this prophecy, but there is no doubt that we are daily nearing its fulfilment.

The general life of the complex being is therefore the more or less perfect synergy, the *ordered process* of elementary lives. General death is the destruction of these partial lives. The nervous system, the instrument of this harmony of the parts, represents the social bond. It keeps most of the partial elements under its sway, and is thus the intermediary of their relations. The closer this dependence, the higher the development of the nervous apparatus, and the better, also, is assured the universal solidarity and therefore the unity of the organism. Cellular federation assumes the characteristic of a unique individuality in proportion to the development of this nervous centralization. With an ideal perfect nervous system the correlation of the parts would also attain perfection. As Cuvier said: "None could experience change without a change in the rest."

But no animal possesses this extreme solidarity of the parts of the living economy. It is a philosopher's dream. It is the dream of Kant, to whom the perfect organism would be "a teleological system," a system of reciprocal ends and means, a sum total of parts each existing for and by the rest, for and by the whole. An organism so completely connected would be unlikely to live. In fact, living organisms show a little more freedom in the interplay of their parts. Their nervous apparatus fortunately does not attain this imaginary perfection; their unity is not so rigorous. The idea of individuality, of individual existence, is therefore not absolute but relative. There are all degrees of it according to the development of the nervous system. What the man in the street and the doctor himself understand by death is the situation created by the stopping of the general wheels, the brain, the heart, and the lungs. If the breath leaves no trace on the glass held to the mouth, if the beating of the heart is no longer perceptible by the hand which touches or the ear which listens, if the movement and the reaction of sensitiveness have ceased to be manifest, these signs make us conclude that it is death. But this conclusion, as we have said before, is a prognostic rather than a judgment of fact. It expresses the belief that the subject will certainly die, and not that it is from this moment dead. To the physiologist the subject is only on the way to die. The process has started. The only real death is when the universal death of all the elements has been consummated.

CHAPTER III.

PHYSICAL AND CHEMICAL CHARACTERS OF CELLULAR DEATH. NECROBIOSIS. GROWING OLD.

Characteristic of elementary life—Changes produced by death in the composition and the death of the cell—Schlemm; Loew; Bokorny; Pflüger; A. Gautier; Duclaux—The processive character of death—Accidental death—Necrobiosis—Atrophy—Degeneration—So-called natural death—Senescence—Metchnikoff's theory of senescence—Objections.

ELEMENTARY death is nothing but the suppression in the anatomical elements of *all* the phenomena of vitality.

Characteristics of Elementary Life.—The characteristic features of elementary life have been sufficiently fixed by science. First of all, there is *morphological unity.* All the living elements have an identical morphological composition. That is to say that life is only accomplished and sustained in all its fulness in organic units possessing the anatomical constitution of the cell, with its cytoplasm and its nucleus, constituted on the classical type. In the second place, there is *chemical unity.* The constituent matter, the matter of which the cell is built up, diverges but little from a chemical type—a proteid complex, with a hexonic nucleus, and from a physical model which is an emulsion of granulous, immiscible liquids, of different viscosities. The third character consists in the possession of a *specific form* acquired, preserved, and repaired by the element. The fourth character, and perhaps the most essential of all, is *the property of growth* or *nutrition* with its consequence, namely, a relation of exchanges with the external medium, exchanges in which oxygen plays considerable part. Finally, there is a last property, that of *reproduction*, which in a certain measure is a necessary consequence of the preceding,—*i.e.*, of growth.

These five vital characters of the elements are most in evidence in cells living in isolation, in microscopical beings formed of a single cell, protophytes and protozoa. But we find them also in the associations formed by the cells among one another—*i.e.*, in ordinary plants and animals, multicellular complexes, called for this reason metaphytes and metazoa. Free or associated, the anatomical elements behave in the same way—feed, grow, breathe, digest in the same manner. As a matter of fact, the grouping of the cells, the relations, proximity and contiguity, which they assume, introduce some variants into the expression of the common phenomena; but these slight differences cannot disguise the essential community of the vital processes.

The majority of physiologists, following Claude Bernard, admit as valent and convincing the proof that the illustrious experimenter furnished of this unity of the vital processes. There are, however, a few voices crying in the wilderness. M. Le Dantec is one. In his new theory of life he amplifies and exalts the differences which exist between the elementary life of the proteids and the associated life of the metazoa. In them he can see nothing but contrasts and deviations.

If this is elementary life, let us ask what is *elementary death—i.e.*, the death of the cell. And in this connection let us ask the questions which we have to examine in the case of animals high in organization, and of man himself. What are the characteristics of elementary death? When the cell dies, is its death preceded by a growing old or senescence? What are the preliminary signs and the acknowledged symptons?

Changes Produced by Death.—The state of death is only truly realized when the fundamental properties of living matter enumerated above have entirely disappeared. We must follow step by step this disappearance in all the anatomical elements of the metazoan.

Now the properties of the cell are connected with the physical and chemical organization of living matter. For them to disappear entirely, this organization must be destroyed as far as all that is essential in it is concerned. We cannot admit with the vitalists that there is any material difference between the dead and the living, and that only an immaterial principle which has escaped into the air distinguishes the corpse from the animated being. In fact, the external configuration may be almost preserved, and the corpse may bear the aspect and the forms of the preceding state. But this appearance is deceptive. Something in reality has changed. The structure, the chemical composition of the living substance, have undergone essential changes. What are these changes?

Physical Changes.—Certain physiologists have endeavoured to determine them. Klemm, a botanist, pointed out in 1895 the physical changes which characterize the death of vegetable cells—loss of turgescence, fragmentation of the protoplasm, the formation of granules, and the appearance of vacuoles.

Chemical Changes.—O. Loew and Bokorny laid great stress in 1886 and 1896 on the chemical changes. The living protoplasm according to them is an unstable proteid compound. A slight change would detach from the albuminoid molecule a nucleus with the function of aldehyde, and at the same time would transform an amido-group into an amido-group. This would suffice for the transition of the protoplasm from the living to the dead state. This theory is based on the fact that the compounds which exercise a toxic action on the living cell, without acting chemically on the dead albumin, are easily fixed by the aldehydes; and on the fact that many of them, which attack simultaneously the living albuminoids and the dead albumin, easily combine with the amido-group.

E. Pflüger, a celebrated German scientist, has considered living matter as an albumin spontaneously decomposable, the essential nucleus of which is formed by cyanogen. Its active instability would be due to the penetration into the molecule of the oxygen which fixes on the carbon and separates it from the nitrogen. Armand Gautier has not confirmed this view. Duclaux (1898) has stated that the difference between the living and the dead albumin would be of a stereo-chemical order.

Progressive Character of Death. Accidental Death.—We have seen that in general the disappearance of the characteristics of vitality is not instantaneous, at least in the natural course of things, in complex organisms. It is the end of a more or less rapid process. But death is not instantaneous in the isolated anatomical element any more than it is in the protozoan or protophyte. We must have recourse to very violent devices of destruction to kill the cell at a blow, to leave absolutely nothing of its organization existing. The protoplasm of yeast when violently crushed by Büchner still possessed the power of secreting soluble ferments. A powerful action, a very high temperature, is necessary to obtain the result. *A fortiori*, the difficulty increases in the case of complex organisms, all of whose living elements cannot be attacked at the same moment by the destructive cause. A mechanical action, capable of destroying at one blow all the living parts of a complex being, of an animal, of a plant, must be of almost inconceivable power. The blow of a Nasmyth hammer would not be strong enough.

The chemical alteration produced by a very toxic substance distributed throughout the blood, and thus brought into contact with each element, would produce a disorganization which, however rapid it were, could not be called instantaneous. And the same holds good of physical agents.

But these are not the processes of nature under normal circumstances. They are accidents or devices. We shall leave on one side their consideration and we shall only deal here with the natural processes of the organism.

Imagine it placed in a medium appropriate to its needs and following out without intervening complications the evolution assigned to it by its constitution. Experiment tells us that this natural evolution in every case known to us ends in death. Death supervenes sooner or later. For beings higher in organization, which we can bring into closer and closer resemblance to man, we find that they die of disease, by accident, or of old age. And as disease is an accident, we may naturally ask if what we call old age is not also a disease.

However that may be, the mortal process, being never instantaneous, has a duration, a beginning, a development, an end—in a word, a history. It constitutes an intermediary phase between perfect life and certain death.

Necrobiosis. Atrophy. Degeneration.—The process according to the circumstances may be shortened or prolonged. When death is the result of violence events are precipitated. The physical and chemical transformations of the living matter constitute a kind of acute alteration called by Schultze and Virchow *necrobiosis.* According to the pathologists, there are two kinds of *necrobiosis*:—that by *destruction*, by *simple atrophy*, which causes the anatomical elements to disappear gradually without undergoing appreciable modifications; and *necrobiosis by degeneration*, which transforms the protoplasm into fatty matter into calcareous matter, into granulations (fatty degeneration, calcification, granulous degeneration). There is no disagreement as to the causes of this necrobiosis. They are always accidental; they originate in external circumstances:—the insufficiency of the alimentary materials, of water, of oxygen; the presence in the medium of real poisons destroying the organized matter; the violent intervention of physical agents, heat, electricity; the reflex on the composition of the cellular atmosphere of a violent attack on some essential organ, the heart, the lungs, the kidneys.

Senescence. Old Age.—In a second category we must place the mortal processes, slow in their movement, in which we cannot see the intervention of clearly accidental and abnormal disturbing agents. Death appears to be the termination of a breaking-up proceeding by insensible degrees in consequence of the progressive accumulation of very small inappreciable perturbations. This slow breaking up is adequately expressed by the term—growing old, or senescence. The alterations by which it is betrayed in the cell are especially *atrophic*, but they are also accompanied, however, by different forms of degeneration. An extremely important question arises on this subject, and that is whether the phenomena of senility have their cause in the cell itself, if they are inevitably found in its organization, and therefore if old age and death are natural and necessary phenomena. Or, on the other hand, should we consider them as due to a progressive alteration of the medium, the character of which would be accidental although frequent or habitual? This, in a word, is the problem which has so often engaged the attention of philosophical biologists. Are old age and death natural and inevitable phenomena?

The recent experiments of Loeb and Calkins, and all similar observations, tend to attribute to the phenomenon of growing old the character of a remediable accident. But the remedy has not been found, and the animal finally succumbs to these slow transformations of its anatomical elements. We then say that it *dies of old age*.

Metchnikoff's Theory of Senescence. Objections.—Metchnikoff has proposed a theory of the mechanism of this general senescence. The elements of the conjunctive tissue, phagocytes, macrophages, which exist everywhere around the specialized and higher anatomical elements would destroy and devour them as soon as their vitality diminishes, and would take their place. In the brain, for example, it would be the phagocytes which, attacking the nervous cellules, would disorganize the higher elements, incapable of defending themselves. This substitution of the conjunctive tissue, which only possesses vegetative properties of a low order, for the nervous tissues, which possesses very high vegetative properties, results in an evident breaking-up. The gross element of violent and energetic vitality stifles the refined and higher element.

This expulsion is a very real fact. It constitutes what is called senile sclerosis. But the active *rôle* attributed to it by Metchnikoff in the process of degeneration is not so certain. An expert observer in the microscopic study of the nervous system, M. Marinesco, does not accept this interpretation as far as the senescence of the elements of the brain is concerned. Diminution of the cell, the decrease in the number of its stainable granulations, chromatolysis, the formation of inert, pigmented substances—all these phenomena which characterize the breaking-up of the cerebral cells would be accomplished, according to this observer, without the intervention of the conjunctive elements, the phagocytes.

The characteristic of extensive and progressive process presented by death necessitates in a complex organism, which is a prey to it, the existence side by side of living and dead cells. Similarly, in the organism which is growing old, there are young elements and elements of every age side by side with senile elements. As long as the disorganization of the last has not gone too far, they may be rejuvenated.

All we have to do is to restore to them an appropriate ambient medium. The whole question is one of knowing and being able to realize, for this or that part which we wish to reanimate and to rejuvenate, the very special or very delicate conditions that this medium must fulfil. As we have said, success is attained in this respect as far as the heart is concerned, and this is why we are able to reanimate and to revive the heart of a dead man. It is hoped that ideas along these lines will extend with the progress of physiology.

After this sketch of the conditions and of the varieties of cellular death we must return to the essential problem which is engaging the curiosity of biologists and philosophers. Is death unavoidable, inevitable? Is it the necessary consequence of life itself, the inevitable issue, the inevitable end?

There are two ways of endeavouring to solve this question of the inevitability of death. The first is to examine popular observation, practised, so to speak, unintelligently and without special precautions. The second is to analyze everything we know relative to the conditions of elementary life.

CHAPTER IV.

THE APPARENT PERENNITY OF COMPLEX INDIVIDUALS.

Millenary trees—Plants with a definite rhizome—Vegetables reproduced by cuttings—Animal colonies—Destruction due to extrinsic causes—Difficulty of interpretation.

POPULAR opinion teaches us that living beings have only a transient existence, and as a poet has said: "Life is but a flash between two dark nights." But, on the other hand, simple observation shows us, or appears to show us, beings whose duration of existence is far longer, and practically illimitable.

Millenary Trees.—We know of trees of venerable antiquity. Among these patriarchs of the vegetable world there is a chestnut tree on Mount Etna which is ten centuries old, and an ivy in Scotland which is said to be thirty centuries old. Trees of 5000 years old are not absolutely unknown. We may mention among those of that age the famous dragon tree[1] at Orotava, in the island of Teneriffe. Two other examples are known in California—the pseudo-cedar, or *Tascodium*, at Sacramento, and a *Sequoïa gigantea.* We know that the olive tree may live 700 years. There are cedars 800 years old and oaks of the age of 1,500 years.

Plants with a Rhizome.—Vegetable species of

[1] Lately destroyed in a storm. [Tr.]

almost unlimited duration of life are known to botanists. Such, for instance, are plants with a definite rhizome, such as colchicum. Autumnal colchicum has a subterranean root, the bulb of which pushes out every year fresh axes for a new bloom; and as each of these new axes stretches out an almost constant length, a botanist once set himself the singular problem of discovering how long it would take such a foot, if suitably directed, to travel round the world.

Vegetables Reproduced by Cuttings.—Vegetables reproduced by slips furnish another example of living beings of indefinite duration. The weeping willows which adorn the banks of sheets of water in the parks and gardens throughout the whole of Europe have sprung, directly or indirectly, from slips of the first *Salix Babylonica* introduced to the West. May it not be said that they are the permanent fragments of that one and the same willow?

Animal Colonies.—These examples, as well as those furnished to zoologists by the consideration of the polypi which have produced by their slow growth the reefs, or *atolls*, of the Polynesian seas, do not, however, prove the perennity of living beings. The argument is valueless, for it is founded upon a confusion. It turns on the difficulty that biologists experience in defining the individual. The oak and the polypus are not simple individuals, but associations of individuals, or, to use Hegel's expression, the nations of which we see the successive generations. We give to this succession of generations a unique existence, and our reasoning comes to this, that we confer on each present citizen of this social body the antiquity which belongs to the whole.

Destruction of the Social Individual due to Extrinsic Causes.—As for the destruction, the death of this social individual, of this hundred-year-old tree, it seems indeed that there is no ground for considering it a natural necessity. We find the sufficient reason of its usual end in the repercussion on the individual of external and contingent circumstances. The cause of the death of a tree, of an oak many centuries old, is to be found in the ambient conditions, and not in some internal condition. Cold and heat, damp and dryness, the weight of the snow, the mechanical action of the rain, of hail, of winds unchained, of lightning; the ravages of insects and parasites—these are what really work its ruin. And further, the new branches, appearing every year and increasing the load the trunk has to bear, increase the pressure of the parts, and make more difficult the motion of the sap. But for these obstacles, external, so to speak, to the vegetable being itself, it would continue indefinitely to bloom, to fructify, and as each spring returned to show fresh buds.

Difficulty of Interpretation.—In this as in all other examples we must know the nature of the beings that we see lasting on and braving the centuries. Is it the individual? Is it the species? Is it a living being, properly so called, having its unity and its individuality, or is it a series of generations succeeding one another in time and extending in space? In a word, the question is one of knowing if we have to do with a real tree or with a genealogical tree. We are just as uncertain when we deal with animals. What is the being that lasts on—a series of generations or an individual? This doubt forbids us to draw any conclusion from the observation of complex beings.

We must therefore return from them to the *elementary being*, and we must examine it from the point of view of perennity or of vital decay. Let us then ask the questions that we have already examined with reference to animals high in organization and to man himself. Is the death of the cell an inevitable characteristic? Are there any cells, protophytes, protozoa, which are immortal?

CHAPTER V.

THE IMMORTALITY OF THE PROTOZOA.

Impossibility of life without evolution—Law of increase and division—Immortality of the protozoa—Death, a phenomenon of adaptation which has appeared in the course of the ages—The infusoria—The death of the infusoria—Two kinds of reproduction—The caryogamic rejuvenescence of Maupas—Calkins on rejuvenescence—Causes of senescence—Impossibility of life without evolution.

WE take into account, *a priori*, the conditions that must be fulfilled by the monocellular being in order to escape the inevitability of evolution, of the succession of ages, of old age, and of death. It must be able indefinitely to maintain itself in a normal régime, without changing, without increasing, maintaining its constant morphological and chemical composition, in an environment vast enough for it to be unaltered by the borrowings or the spendings resulting from its nutrition—*i.e.*, it must remain constant in the presence of the constant being. We might conceive of a nutrition perfect enough, of exchanges exact enough, and regular enough, for the state of things to be indefinitely maintained. This would be absolute permanence realized in the vital mobility.

The Law of Growth and Division.—This model of a perfect and invariable machine does not exist in nature. Life is incompatible with the absolute permanence of the dimensions and the forms of the living organism.

In a word, it is a rigorous law of living nature that the cell can neither live indefinitely without growth, nor grow indefinitely without division.

Why is this so? Why is there this impossibility of a regular régime in which the cell would be maintained in magnitude without diminution or increase? Why has nutrition as a necessary consequence the growth of the element? This is what we do not positively know.

Things are so. It is an irreducible fact, peculiar to the protoplasm, a characteristic of the living matter of the cell. It is the fundamental basis of the property of generation. That is all we can say about it. Real living beings have therefore inevitably an evolution. They are not unchangeable. In its simple form this evolution consists in the fact that the cell grows, divides, and diminishes by this division, begins the upward march which ends in a new division. And so on.

Immortality of the Protozoa.—It may happen, and it does happen in fact, that this series of acts is repeated indefinitely at any rate unless an accidental cause should interrupt it. The animal thus describes an indefinite curve, constituted by a series of indentations, the highest point of which corresponds to the maximum of size, and the lowest point to the diminution which succeeds the division. This state of things has no inevitable end if the medium does not change. The being is immortal.

In fact, the compound beings of a single cell, protophytes and protozoa, the algae and the unicellular mushrooms, at the minimum stage of differentiation, escape the necessity of death. They have not, as Weismann remarks, the real immortality of the gods of mythology, who were invulnerable. On the contrary, they are infinitely vulnerable, fragile, and perishable; myriads die every moment. But their death is not inevitable. They succumb to accidents, never to old age.

Imagine one of these beings placed in a culture medium favourable to the full exercise of its activities, and, moreover, wide enough in its extent to be unaffected by the infinitely small quantities of material which the animal may take from it or expel into it. Suppose, for example, it is an infusorian in an ocean. In this invariable medium the being lives, increases, and grows continually. When it has reached the limits of a size fixed by its specific law, it divides into two parts, which are indistinguishable the one from the other. It leaves one of its halves to colonize in its neighbourhood, and it begins its evolution as before. There is no reason why the fact should not be repeated indefinitely, since nothing is changed, either in the medium or in the animal.

To sum up. The phenomena which take place in the cell of the protozoan do not behave as a cause of check. The medium allows the organism to revictual and to discharge itself in such a way and with such perfection that the animal is always living in a regular régime, and, with the exception of its growth and later on of its division, there is nothing changed in it.

Death a Phenomenon of Adaptation—It appeared in the Course of the Ages.—This immortality belongs in principle to all the protista which are reproduced by simple and equal division. If it be remarked that these rudimentary organisms endowed with perennity are the first living forms which have shown themselves on the surface of the globe, and that they have no doubt preceded many others—the multicellular, for instance, which are liable, on the contrary, to decay—the conclusion is obvious:—Life has long existed without death. Death has been a phenomenon of adaptation which has appeared in the course of the ages in consequence of the evolution of species.

The Death of Infusoria.—We may ask ourselves at what moment in the history of the globe, at what period of the evolution of its fauna, this novelty, death, made its appearance. The celebrated experiments of Maupas on the senescence of the infusoria seem to authorize us to give a precise answer to this question. By means of these experiments we are led to believe that death must have appeared at the same time as sexual reproduction. Death became possible when this process of generation was established, not in all its plenitude, but in its humblest beginnings, under the rudimentary forms of unequal division and of conjugation. This happened when the infusoria began to people the waters.

The Two Modes of Multiplication.—Infusoria are, in fact, capable of multiplication by simple division. It is true to say that in addition to this resource, the only one which interests us here, because it is the only one which confers immortality, they possess another. They present and exercise under certain circumstances a second mode of reproduction, caryogamic conjugation. It is a rather complicated process in its detail, but it is definitively summed up as the temporary pairing of two individuals, which are otherwise very much alike, and which cannot be distinguished as male and female. They become closely united on one of their faces; they reciprocally exchange a semi-nucleus which passes into the conjoint individual; and then they separate. But infusoria can be prevented from this conjunction by regularly isolating them immediately after their birth. Then they grow, and are constrained after a lapse of time to divide according to the first method.

Maupas has shown that the infusoria could not accommodate themselves to this régime indefinitely; they could not go on dividing for ever. After a certain number of divisions they show signs of degeneration and of evident decay. The size diminishes, the nuclear organs become atrophied, all the activities fail, and the infusorian perishes. It succumbs to this kind of senile atrophy unless it is given an opportunity of conjugation with another infusorian in the same plight. In this act it then derives new strength, it grows larger, attains its proper size, and builds up its organs once more. Conjugation gives it life, youth, and immortality.

Alimentary Rejuvenescence.—Recent observations due to Mr. G. N. Calkins, an American biologist, and confirmed by other investigators, have shown that this method of rejuvenescence is not the only one, and is not even the most efficacious. Conjugation has no mysterious, specific virtue. The infusoria need not be married in order to be rejuvenated. It is sufficient to improve their food. In the case of the "tailed" paramecium we may substitute beef broth and phosphates for conjugation. Calkins observed 665 consecutive generations without blemish, without exhaustion, and without any sign of old age. Plenty of food and simple drugs have successfully resisted senility and the train of atrophic degenerations which it involves.

Causes of Senescence.—As for the causes of senescence which have been remedied with such success, they are not exactly known. Calkins thinks that senescence results from the progressive losses to the organism of some substance essential to life. Conjugation or intensive alimentation would act by building up again this necessary compound. G. Loisel believes on the contrary that it is a matter of the progressive accumulation of toxic products due to a kind of alimentary auto-intoxication.

CHAPTER VI.

LETHALITY OF THE METAZOA AND OF DIFFERENTIATED CELLS.

Evolution and death of metazoa.—Possible rejuvenescence of the differentiated cells by the conditions of the medium. — Conditions of the medium for immortal cells. — The immortal elements of metazoa.—The element in accidental and remediable death.—Somatic cells and sexual cells.

Evolution and Death of Metazoa.—We have seen that the infusoria are no longer animals in which material exchanges take place with sufficient perfection, and in which cellular division, the consequence of growth, is produced with sufficient precision and equality for life to be carried on indefinitely in a perfect equilibrium in the appropriate medium without alteration or check. *A fortiori* we no longer find the perfect regularity of nutritive exchange in the classes above them. In a word, starting from this inferior group, there are no animated beings in the state of existence which Le Dantec calls "condition 1°" of manifested life?" Living matter, instead of being continually kept identical in conditions of identical media, is modified in the course of existence. It becomes dependent on time. It describes a declining trajectory; it experiences evolution, decay, and death. Thus the fundamental condition of invariable youth and of immortality fails in all metazoa. The vital wastes accumulate in all through the insufficiency or the imperfection of nutritive absorption or of excretion. Life decays; the organism progressively alters, and thus is constituted that state of decrepitude by atrophy or chemical modification which we call senescence, and which ends in death. To sum up, old age and death may be attributed to cellular differentiation.

Possible Alimentary Rejuvenescence of the Differentiated Cells—Conditions of Medium.—We must add, however—as the teaching of experiments in general and in particular as the teaching of the experiments of Loeb and of Calkins—that a slight change of the environment, made at the right time, is capable of re-establishing equilibrium and of completely rejuvenating the infusorian. Senescence has not in this case a definitive any more than an intrinsic character; a modification in the composition of the alimentary medium will successfully resist it. If we are allowed to generalize this result, it may be said that senescence, the declining trajectory, the evolution step by step down to death, are not for the cells considered in isolation an inevitable and essentially inherent in the organism, and a rigorous consequence of life itself. They preserve an accidental character. In senescence and death there is no really natural, internal cause, inexorable, and irremediable, as was claimed in the past by J. Müller, and more recently by Cohnheim in Germany and Sedgwick Minot in America.

Conditions of the Medium for Immortal Cells.—As for the cells which are less differentiated, the protophytes and the protozoa situated one degree lower in the scale than the infusoria, we must admit the possibility of that perfect and continuous equilibrium which would save them from senile decrepitude. And it is quite understood that this privilege remains subordinated to the perfect constancy of the appropriate medium. If the latter changes, the equilibrium is broken, the small insensible perturbations of nutrition accumulate, vital activity decays, and in sole consequence of the imperfection of the extrinsic conditions or of the medium, the living being finds itself once more dragged down to decay and to death.

Immortal Elements of the Metazoa.—All the preceding facts and considerations refer to isolated cells, to monocellular beings. But, and this is what makes these truths so interesting, they may be extended to all cells grouped in collectivity—*i.e.*, to all the animals and living beings that we know. In the complicated edifice of the organism, the anatomical elements, at any rate the least differentiated, would have a continual brevet of immortality. Generally speaking, this would be the case for the egg, for the sexual elements, and perhaps, too, for the white globules of the blood, the leucocytes. And, further, around each of these elements must be realized the invariably perfect medium which is the necessary condition. This does not take place.

Elements in Accidental and Remediable Death.—As for the other elements, they are like the infusoria, but without the resource of conjugation. The ambient medium becomes exhausted and intoxicated around each cell, in consequence of the accidents which happen to the other cells. Each therefore undergoes progressive decay, and finally they perish—the decay and destruction being perhaps in principle accidental, but, in fact, they are the rule.

The different anatomical elements of the organism are more or less sensitive to those perturbations which cause senescence, necrobiosis, and death. There are some more fragile and more exposed. Some are more resisting, and finally, there are some which are really immortal. We have just said that the sexual cell, the ovum, is one. It follows that the metazoan, man for instance, cannot entirely die. Let us consider one of these beings. Its ancestors, so to speak, have not entirely disappeared; each has left the fertile egg, the surviving element from which has issued the being of which we speak; and when it in its turn has developed, part of that ovum has been placed in reserve for a new generation. The death of the elements is not therefore universal. The metazoan is divided from the beginning into two parts. On the one hand are the cells destined to form the body, *somatic* cells. They will die. On the other hand are the *reproductive*, or *germinal*, or *sexual* cells, capable of living indefinitely.

Somatic and Sexual Cells.—In this sense we may say with Weismann that there are two things in the animal and in man—the one mortal, the *soma* the body, the other immortal, the *germen*. These germinal cells, as in the case of the protozoa we mentioned above, possess a conditional immortality. They are imperishable, but on the contrary, are fragile and vulnerable. Millions of ova are destroyed and are disappearing every moment. They may die by accident, but never of old age.

We now understand that if the protistae are immortal, it is because these living beings, reduced to a single cell, accumulate in it the compound characters of the somatic cell and germinal cell, and enjoy the privilege which is attached to the latter.

CHAPTER VII.

MAN. THE INSTINCT OF LIFE AND THE INSTINCT OF DEATH.

The miseries of humanity: 1. Disease; 2. Old age.—Old age considered as a chronic disease.—Its occasional cause.—3. The disharmonies of human nature; 4. The instinct of life and the instinct of death.

MAN'S unhappy plight is the constant theme of philosophies and religions. Without referring to its moral basis, it has a physical basis due to four causes—the physical imperfection or disharmony of nature, disease, old age, and death—or rather of three, for what we call old age is perhaps a simple disease. These are the great sorrows of man, the sources of all his woes. Disease attacks him, old age awaits him, and death must tear him from all the ties which he has formed. All his pleasures are poisoned by the certain knowledge that they last but for a moment, that they are as precarious as his health, his youth, and his life itself.

§ DISEASE.

Disease, frequent, constant, and inevitable as it is, is, however, nothing but a fact outside the natural order. Its character is clearly accidental, and it interrupts the normal cycle of evolution. Medical observation teaches us, on the other hand, that the health of the body reacts on that of the mind; and therefore man as a whole, moral and physical, is affected by disease. Bacon described a diseased body as a jailer to the soul, and the healthy body as a host. Pascal recognized in diseases a principle of error. "They spoil our judgment and our senses."

I am not expressing a chimerical hope when I predict that science will conquer disease. Medicine has at last issued from the contemplative attitude of so many centuries; it has engaged in the struggle, and signs of victory are already appearing. Disease is no longer the mysterious power which it was impossible to escape. Pasteur gave to it a body. The microbe can be caught. In the words of Schopenhauer, an alteration of the atmosphere so slight that it is impossible to detect it by chemical analysis may bring on cholera, yellow fever, the black plague, diseases which carry off thousands of men; and a slightly greater alteration might endanger all life. The at once mysterious and terrifying spectacle of the cholera at Berlin in 1831 had such an effect on the philosopher that he fled in terror to Frankfort. It has been said that this was the origin of his pessimism, and that but for this he would have continued to teach idealistic philosophy in some Prussian university. L. Hartmann, another celebrated leader of contemporary pessimism, has also said that disease will always be beyond the resources of medicine. Facts have given the lie to these sombre prognostics. The microbic origin of most infectious diseases has been recognized. The discovery of attenuated poisons and serums has diminished their gravity. An exact knowledge of methods of contagion has enabled us to erect against them impregnable barriers. Cholera, yellow fever, the plague knock in vain at our doors. Diphtheria, dreaded by every mother, has partially lost its deadly character. Puerperal fever and blindness of the new-born child are tending to disappear. Legend tells us that Buddha in his youth, frightened at the sight of a sick man, expressed in his father's presence the wish to be always in perfect health and sheltered from disease. The King answered: "My son! you are asking the impossible." But it is towards the realization of this impossibility that we are on our way. Science is repelling the attacks of disease.

§ 2. Old Age.

Old age is another sorrow of humanity. The stage of existence in which the strength grows less and never grows greater, and in which a thousand infirmities appear, is not, however, a stage universal in animals. Most of them die without our perceiving in them any apparent signs of senile weakness. On the other hand, some vegetables exhibit these signs. Some trees are old; but it is in birds and mammals that this decay, with the train of evils which accompanies it, becomes a very marked phase of existence. In man to debility is added a bodily shrinkage, grey hairs, withered skin, and the wearing out and loss of teeth. The exhausted and atrophied organism offers a favourable field to all intercurrent diseases and to every cause of destruction. It is this discrepitude which makes old age so hateful. All desire to be old, said Cicero; and when they are old, they say that old age has come quicker than they expected.

La Bruyère expresses it in an apothegm, "We want to grow old, and we fear old age." One would like longevity without old age.

But can life be prolonged without senility diminishing its value? Metchnikoff thinks it can. He more or less clearly catches a glimpse of a normal evolution of existence which would make it longer and nevertheless exempt from senile decay.

It is remarkable that we have so few scientific data on the old age of man, and we have still fewer on that of animals. The biologist knows no more than the layman. The old age of the dog is betrayed by its gait. Its coat loses its lustre, just as in disease. The hair whitens around the forehead and the muzzle. The teeth grow blunt and drop out. The character loses its gaiety and becomes gloomy; the animal becomes indifferent. He ceases to bark, and often becomes blind and deaf.

It is admitted that senile degeneration is due to an alteration affecting most of the tissues. The cells, the special anatomical elements of the liver, the kidney, and the brain are reduced by atrophy and degeneration. At the same time, the conjunctive woof which serves them as a support develops, on the contrary, at the expense in a measure of the higher elements. For this reason the tissues harden. We know that the flesh of old animals is tough. We know in pathology that this is happening to the tissues. It is due to growth, to injury to the active and important elements, to the elements of support of the organs. They form a tissue sometimes called packed tissue, to show its secondary rôle with reference to the elements which are deposited in it. This kind of degeneration of the organs is known as sclerosis. It constitutes the charasteristic lesion of a certain number of chronic diseases; and these diseases are serious, for the stifling of the characteristic elements by the less important elements of the conjunctive or packed tissue results in the more or less complete reduction or suppression of the function.

The blood vessels also undergo this transformation, and what we may call universal trouble and danger ensue. This sclerosis of the arteries, this arterio-sclerosis, not only deprives the walls of the blood vessels of the suppleness and elasticity which are necessary for the proper irrigation of the organs, but it makes them more fragile. Thus it becomes a cause of hemorrhage, which is a very serious matter as far as the brain and lungs are concerned.

It is remarkable that the alteration of the tissues during old age should be exactly similar to this. This is inferred from the few researches that have been made on the subject—from those of Demange in 1886, of Merkel in 1891, and finally from the researches of Metchnikoff himself. It is a generalized sclerosis. As its consequence we have the lowering of the proper activity of the organs and the danger of cerebral hemorrhage created by arterio-sclerosis. The transformations of the tissues in old men are therefore summed up in the atrophy of the important and specific elements of the tissues, and their replacement by the hypertrophied conjunctive tissue. This sclerosis is comparable to that of chronic diseases; it is a pathological condition. Thus old age, as we understand it, is a chronic disease and not a normal phase of the vital cycle.

On the other hand, if we ask ourselves what is the

origin of the scleroses which engender chronic diseases, we find that they are due to the action of various poisons, among which syphilitic poison and the immoderate use of alcohol take the first place. These are also the usual causes of senile degeneration. But there must be some other, some very general cause to explain the universality of the process of senescence. Metchnikoff thinks that he has found this cause in the microbes which swarm in man's digestive tube, particularly in the large intestine. Their number is enormous. Strassburger has given an approximate calculation, but words fail to express it. We have to imagine a figure followed by fifteen zeros. This microbic flora is composed of "bacilli" and of "cocci," and comprises a third of the rejected matter. It produces slow poisons, which, being at once reabsorbed, pass into the blood and provoke the constant irritation from which results arterio-sclerosis and the universal sclerosis of old age. Instead of enjoying a healthy and normal old age, in which the faculties of ripening years are preserved, we drag out a diminished life, a kind of chronic disease, which is ordinary old age. This is due, according to Metchnikoff, to the parasitism and the symbiosis of microbic flora, lodged in a part of the economy in which it finds all the conditions favourable to its prolific expansion. Such is the specious theory, held to the verge of intrepidity, by which this investigator explains the misery of our old age, and which inspires him with the idea of a remedy. For his observations conclude with a régime, a series of prescriptions by which the author fancies that life may be lengthened and the evils of old age swept from our path. The dangerous flora must be transformed into a cultivated and selected flora. Although the organ in question may be of doubtful utility, and although its existence, the legacy of atavic heredity, must be considered as a disharmony of human nature, Metchnikoff does not go so far as to propose that it should be cut away, and that we should call in surgery to assist in making mankind perfect! But the rational means he proposes will be endorsed by the most judicious students of hygiene; and their effect, if it not as wonderful as one hopes for, cannot fail to ameliorate the conditions of old age and make it more vigorous.

§ 3. Disharmonies in Human Nature.

Another misery in the condition of man is due to the dissidencies of his nature—that is to say, to his physical imperfections and the discordancies which exist between the physiological functions and the instincts which should regulate them.

This discordance reigns throughout the physical organism. The body of man is not the perfect masterpiece it was once supposed to be. It is encumbered with annoying inutilities, with rudimentary organs that have neither rôle nor function, unfinished sketches which nature has left in the different parts of his body. Such are the lachrymal caruncle, a vestige of the third eyebrow in mammals; the extrinsic muscles of the ear; the pineal gland of the brain, which is only the rudiment of an ancestral organ; the third eye, or the Cyclopean eye of the saurians. The list is interminable. Wiedersheim has counted in man 107 of these abortive hereditary organs, the useless vestiges of organs useful to our remote animal ancestors, atrophied in the course of ages in consequence of modifications that have taken place in the external medium.

These rudimentary organs are not only useless; they are often positively harmful.

But the most serious discordance is that which exists between the physiological functions and the instincts which regulate them. In a well-regulated organism slowly developed by adaptation the instincts and the organs alike should be in relation with the functions. All really natural acts are solicited by an instinct, the satisfaction of which is at once a need and a pleasure. The maternal instinct is awakened at the proper moment in animals, and it disappears as soon as the offspring requires no more assistance. A craving for milk is shown in all new-born children, and often disappears at an early age.

Nature has endowed man as well as the other animals with peculiar instincts, destined to preside over the different functions and to ensure their accomplishment. And, at the same time, it has enabled him in a measure to deceive those instincts and to satisfy them by other means than the execution of the physiological acts with a view to which they exist. Love and the instinct of reproduction exist in man before the age of puberty. Canova felt the spur of love at the age of five. Dante was in love with Beatrice at nine; and Byron, then scarcely seven, was already in love with Maria Duff. On the other hand, puberty has no necessary relation to the general maturity of the organism.

The family instinct is subject to the same aberrations. Man limits the number of his children. The

Turks of to-day follow the ancient Greeks in the practice of abortion. Plato approved of the custom, and Aristotle sanctioned its general prevalence. In the province of Canton the Chinese of the agricultural classes kill two-thirds of their girl children, and the same is done at Tahiti. All these customs co-exist with the perfect love and tender care of the living children.

Because of these different discordancies the physical life of man is insufficiently regulated by nature. Neither the physiological instinct, nor the family instinct, nor the social instinct is, in general, sufficiently imperative and precise. Hence, since the internal impulse has not sufficient power, the necessity arises for a rule of conduct exercising its influence from without. Philosophies, religions, and legislation have provided for this. They have regulated man's hygiene and the carrying out of his different physiological functions. Their control has, moreover, had its hygienic side. The scientific hygiene of to-day has inherited their rôle.

The idea of the fundamental perversity of human nature is born of our cognizance of its discordancies, unduly amplified and exaggerated. Soul and body have been considered as distinctly discordant and hostile elements. The body, the shroud of the soul, the temporary host, the prison, the present source of miseries, has been subjected to every kind of mortification. Asceticism has treated the body and all the innate instincts as our mortal foes.

This suspicion, this depreciation of human nature was the great error of the mystics. This view was as fatal as the inverse view of pagan antiquity. The model of the perfect life according to Greek philosophy is a life in conformity with nature. To aim at the harmonious development of man was the precept of the ancient Academy, formulated by Plato. The Stoics and the Epicureans had adopted the same principle. Physical nature is considered as good. It gives us the type, the rule, and the measure. The moral rule itself is exactly appropriate to the physical nature. We may say that pagan morality was hygiene, the hygiene of the soul and the body alike; the *mens sana in corpore sano* gave individual and social direction. The Rationalists, the philosophers of the eighteenth century, such as Baron d'Holbach and later W. Von Humboldt, Darwin, and Herbert Spencer, have adopted analogous views. If these views have been contested, it is because of the imperfections or aberrations of the natural instincts of man. Also, if we wish to base individual family or social morality on the natural instincts of man, it must be specified that these instincts are to be regularized. We must necessarily appeal from the imperfect instincts of the present to the perfected instincts of the future. Their perfection, moreover, will only be a more exact approximation to the real nature of man, and he, having avoided by the aid of science the accidents which cause disease and senile decrepitude, will enjoy a healthy youth and an ideal old age.

The reason of the discrepancies between instinct and function in man is given by the natural history of his development. We know that man has within him original sin—his long atavism. He has sprung, according to the transformists, from a simian stock. He is a cousin, the successful relation, of a type of antinomorphic monkeys, the chimpanzees. He has "arrived," they have remained undeveloped. Probably he had a common ancestor with them, some dryopithecan of an extinct species. From that type sprang a new type already on the way to progress, the *Pithecanthropus erectus*. Finally, the anthropoid ancestor became one fine day the father of a scion, clearly superior to himself, a miraculously gifted being, man. Here, then, is no sign of the slow evolution and gradual progress, which is the doctrine held at present by Transformists. The Dutch botanist De-Vries has shown us, in fact, that nature does leap: *natura facit saltus*. There would thus be crises, as it were, in the life of species. At certain critical epochs considerable differences of a specific value appear in their offspring. It is at one of these critical periods in the simian life that man has appeared as the phenomenal child of an anthropoid. He was born with a brain and an intellect superior to those of his humble parents; and on the other hand, he has inherited from them an organization which is only inadequately adapted to the new conditions of existence created by the development of his sensitiveness and his brain power. This intellect is not proportioned to his organization, which has not developed at the same rate; it protests against the discordances which adaptation has not yet had time to efface. But it will efface them in the future.

§ The Instinct of Life and the Instinct of Death.

The greatest discrepancy of this kind is the knowledge of inevitable death without the instinct which makes it longed for.

There are immortal animals. Man is not of the number. He belongs, like all highly organized beings, to the class of beings which have an end. They die from accident or from disease. They perish in the struggle with other animals, or with microbes, or with external conditions. There are certainly very few, if there are any, which die a really natural death. And so it is with man. We see old men gradually declining who appear to doze gently off into the last sleep, and become extinguished without disease, like a lamp whose oil is exhausted. But this is in most cases only apparently so. Besides the fact that the old age to which they seemed to succumb is really a disease, a generalized sclerosis, autopsy always reveals some lesion more or less directly responsible for the fatal issue.

Man, like all the higher animals, is therefore subject to the law of lethality. But while animals have no idea of death and are not tormented by the sentiment of their inevitable end, man knows and understands this destiny. He has with the animals the instinct of self-preservation, the instinct of life, and at the same time the knowledge and the fear of death. This contradiction, this discordance, is one of the sources of his woes.

Whether it be an accident or the regular term of the normal cycle, death always comes too soon. It surprises the man at a time when he has not yet completed his physiological evolution; hence the aversion and the terror it inspires. "We cannot fix our eyes on the sun or on death," said La Rochefoucauld. The old man does not regard death with less aversion than the young man. "He who is most like the dead dies with most regret." Man knows that he is not getting his full measure.

Further, all the really natural acts are solicited by an instinct, the satisfaction of which is a need and a joy. The need of death should therefore appear at the end of life, just as the need of sleep appears at the end of the day. It would appear, no doubt, if the normal cycle of existence were fulfilled, and if the harmonious evolution were not always interrupted by accident. Death would then be welcomed and longed for. It would lose its horror. The instinct of death would replace at the wished for moment the instinct of life. Man would pass from the banquet of life with no other desire. He would die without regret, "being old and full of days," according to the expression used in the Bible in the case of Abraham, Isaac, and Jacob. No doubt there are some analogies to this in the insects which only assume the perfect form for the purpose of procreation and immediately perish in their full perfection. In these animals the approach of death is blended with the intoxication of hymen. Thus we see some of them, the ephemerae, lose at that moment the instinct of life and the instinct of self-preservation. They allow themselves to be approached, taken, and seized, and make no effort at flight.

But what is this full measure of life which is imparted to us? Metchnikoff holds that the ages attributed to several persons in the Bible are very probable. Abraham lived 175 years, Ishmael 137, Joseph 110, Moses 120. Buffon believed in the existence of a ratio between the longevity of animals and the duration of their growth. He fixed it at

7 : 1. The animal whose development lasts two years would thus have 14 years of life. This law would give us 140 years, but the figure is too high, and Flourens has reduced the ratio to that of 5 : 1, which would still give us 120 years. Plato died in the act of conversation at 81; Isocrates wrote his *Panathenaïcus* at 94; Gorgias died in the full possession of his intellect at 107.

To reach the end of the promised longevity we must neither count on the elixir of life nor on the potable gold of the alchemists, nor on the stone of immortality which did not prevent its inventor, Paracelsus, from dying at the age of 58, nor on transfusion, nor on Graham's celestial bed, nor on King David's gerocomy, nor on any nostrum or remedy. *Contra vim mortis non est medicamen in hortis*, said the Salernian school. What Feuchtersleben said is most true, "The art of prolonging life consists in not cutting it short," and it is a hygiene, but a brilliant hygiene, such as that of which Metchnikoff traces us the future lines, which will realize the desires of nature.

And now shall we find that physiology has solved the enigma proposed by the Sphinx, and that it has answered these poignant questions:—Whence do we come? whither do we go? what is the end of life? The end of life is, to the physiologist as well as to Herbert Spencer, the tendency towards an existence as full and as long as possible, towards a life in conformity with real nature freed from the discordancies which still remain; it is the accomplishment of the harmonious cycle of our normal evolution. This ideal human nature, without discordancies, no longer vitiated as it is at present but improved, will be the work of time and science. Realized at last it will serve as a solid basis for individual, family, and social morality. Healthy youth fit for action; prolonged, adult age, the symbol of strength; normal old age, wise in council, these would have their natural places in harmonious society. "Great actions," said one of old, "are not achieved by exertions of strength, or speed, or agility, but rather by the prudence, the authority, and the judgment which are found in a higher degree in old age." The old age of which Cicero here speaks is the ideal old age, regular and normal, and not the premature, deformed, incapable and egoistic old age which results from a pathological condition. At the end of this full life, the old man being full of days, will crave for the eternal sleep and will resign himself to it with joy. . . .

Death, then, "the last enemy that shall be destroyed," to use the expression of St. Paul, will yield to the power of science. Instead of being "the king of terrors," it will become after a long and healthy life, after a life exempt from morbid accidents, a natural and longed for event, a satisfied need. Then will be realized the wish of the fabulist:—

"*I should like to leave life at this age, just as one leaves a banquet, thanking the host, and departing.*"

Has this physiological solution of the problem of death the virtue attributed to it by Metchnikoff? Is it as optimistic as he thinks it is? The instinct of death supervening at the end of a normal and well-filled cycle will no doubt facilitate to the aged their departure on the great voyage. The wrench will no longer exist for the dead. Will it not exist for those who are left behind? And since the instinct of death can only exist about the time at which death is expected, will the young man and the man of ripened years look with less horror than to-day at the law which cannot be escaped, when they are in full possession of the instinct of life, but warned of the inevitability of death?

INDEX OF AUTHORS.

INDEX OF SUBJECTS.

THE WALTER SCOTT PUBLISHING COMPANY, LTD., FELLING-ON-TYNE.

Life of Bunyan. By Canon Venables.

"A most intelligent, appreciative, and valuable memoir."—*Scotsman.*

Life of Burns. By Professor Blackie.

"The editor certainly made a hit when he persuaded Blackie to write about Burns."—*Pall Mall Gazette.*

Life of Byron. By Hon. Roden Noel.

"He [Mr. Noel] has at any rate given to the world the most credible and comprehensible portrait of the poet ever drawn with pen and ink."—*Manchester Examiner.*

Life of Thomas Carlyle. By R. Garnett, LL.D.

"This is an admirable book. Nothing could be more felicitous and fairer than the way in which he takes us through Carlyle's life and works." —*Pall Mall Gazette.*

Life of Cervantes. By H. E. Watts.

"Let us rather say that no volume of this series, nor, so far as we can recollect, of any of the other numerous similar series, presents the facts of the subject in a more workmanlike style, or with more exhaustive knowledge."—*Manchester Guardian.*

Life of Coleridge. By Hall Caine.

"Brief and vigorous, written throughout with spirit and great literary skill."—*Scotsman.*

Life of Congreve. By Edmund Gosse.

"Mr. Gosse has written an admirable and most interesting biography of a man of letters who is of particular interest to other men of letters." —*The Academy.*

Life of Crabbe. By T. E. Kebbel.

"No English poet since Shakespeare has observed certain aspects of nature and of human life more closely; and in the qualities of manliness and of sincerity he is surpassed by none. . . . Mr. Kebbel's monograph is worthy of the subject."—*Athenæum.*

Life of Darwin. By G. T. Bettany.

"Mr. G. T. Bettany's *Life of Darwin* is a sound and conscientious work."—*Saturday Review.*

Life of Dickens. By Frank T. Marzials.

"Notwithstanding the mass of matter that has been printed relating to Dickens and his works, . . . we should, until we came across this volume, have been at a loss to recommend any popular life of England's most popular novelist as being really satisfactory. The difficulty is removed by Mr. Marzials' little book."—*Athenæum.*

Life of George Eliot. By Oscar Browning.

"We are thankful for this interesting addition to our knowledge of the great novelist."—*Literary World.*

Life of Emerson. By Richard Garnett, LL.D.

"As to the larger section of the public, to whom the series of Great Writers is addressed, no record of Emerson's life and work could be more desirable, both in breadth of treatment and lucidity of style, than Dr. Garnett's."—*Saturday Review.*

Life of Goethe. By James Sime.

"Mr. James Sime's competence as a biographer of Goethe, both in respect of knowledge of his special subject, and of German literature generally, is beyond question."—*Manchester Guardian.*

Life of Goldsmith. By Austin Dobson.

"The story of his literary and social life in London, with all its humorous and pathetic vicissitudes, is here retold as none could tell it better."—*Daily News.*

Life of Nathaniel Hawthorne. By Moncure Conway.

"Easy and conversational as the tone is throughout, no important fact is omitted, no useless fact is recalled."—*Speaker.*

Life of Heine. By William Sharp.

"This is an admirable monograph, . . . more fully written up to the level of recent knowledge and criticism of its theme than any other English work."—*Scotsman.*

Life of Victor Hugo. By Frank T. Marzials.

"Mr. Marzials' volume presents to us, in a more handy form than any English, or even French, handbook gives, the summary of what, up to the moment in which we write, is known or conjectured about the life of the great poet."—*Saturday Review.*

Life of Hunt. By Cosmo Monkhouse.

"Mr. Monkhouse has brought together and skilfully set in order much widely scattered material."—*Athenæum.*

Life of Samuel Johnson. By Colonel F. Grant.

"Colonel Grant has performed his task with diligence, sound judgment, good taste, and accuracy."—*Illustrated London News.*

Life of Keats. By W. M. Rossetti.

"Valuable for the ample information which it contains."—*Cambridge Independent.*

Life of Lessing. By T. W. Rolleston.

"A picture of Lessing which is vivid and truthful, and has enough of detail for all ordinary purposes."—*Nation* (New York).

New York: CHARLES SCRIBNER'S SONS.

Life of Longfellow. By Prof. Eric S. Robertson.

"A most readable little book."—*Liverpool Mercury.*

Life of Marryat. By David Hannay.

"What Mr. Hannay had to do—give a craftsman-like account of a great craftsman who has been almost incomprehensibly undervalued—could hardly have been done better than in this little volume."—*Manchester Guardian.*

Life of Mill. By W. L. Courtney.

"A most sympathetic and discriminating memoir."—*Glasgow Herald.*

Life of Milton. By Richard Garnett, LL.D.

"Within equal compass the life-story of the great poet of Puritanism has never been more charmingly or adequately told."—*Scottish Leader.*

Life of Renan. By Francis Espinasse.

"Sufficiently full in details to give us a living picture of the great scholar, . . . and never tiresome or dull."—*Westminster Review.*

Life of Dante Gabriel Rossetti. By J. Knight.

"Mr. Knight's picture of the great poet and painter is the fullest and best yet presented to the public."—*The Graphic.*

Life of Schiller. By Henry W. Nevinson.

"This is a well-written little volume, which presents the leading facts of the poet's life in a neatly rounded picture."—*Scotsman.*

"Mr. Nevinson has added much to the charm of his book by his spirited translations, which give excellently both the ring and sense of the original."—*Manchester Guardian.*

Life of Arthur Schopenhauer. By William Wallace.

"The series of Great Writers has hardly had a contribution of more marked and peculiar excellence than the book which the Whyte Professor of Moral Philosophy at Oxford has written for it on the attractive and still (in England) little-known subject of Schopenhauer."—*Manchester Guardian.*

Life of Scott. By Professor Yonge.

"For readers and lovers of the poems and novels of Sir Walter Scott this is a most enjoyable book."—*Aberdeen Free Press.*

Life of Shelley. By William Sharp

"The criticisms . . . entitle this capital monograph to be ranked with the best biographies of Shelley."—*Westminster Review.*

New York: CHARLES SCRIBNER'S SONS.

Life of Sheridan. By Lloyd Sanders.

"To say that Mr. Lloyd Sanders, in this volume, has produced the best existing memoir of Sheridan is really to award much fainter praise than the book deserves."—*Manchester Guardian.*

"Rapid and workmanlike in style, the author has evidently a good practical knowledge of the stage of Sheridan's day."—*Saturday Review.*

Life of Adam Smith. By R. B. Haldane, M.P.

"Written with a perspicuity seldom exemplified when dealing with economic science."—*Scotsman.*

"Mr. Haldane's handling of his subject impresses us as that of a man who well understands his theme, and who knows how to elucidate it."—*Scottish Leader.*

"A beginner in political economy might easily do worse than take Mr. Haldane's book as his first text-book."—*Graphic.*

Life of Smollett. By David Hannay.

"A capital record of a writer who still remains one of the great masters of the English novel."—*Saturday Review.*

"Mr. Hannay is excellently equipped for writing the life of Smollett. As a specialist on the history of the eighteenth century navy, he is at a great advantage in handling works so full of the sea and sailors as Smollett's three principal novels. Moreover, he has a complete acquaintance with the Spanish romancers, from whom Smollett drew so much of his inspiration. His criticism is generally acute and discriminating; and his narrative is well arranged, compact, and accurate."—*St. James's Gazette.*

Life of Thackeray. By Herman Merivale and Frank T. Marzials.

"The book, with its excellent bibliography, is one which neither the student nor the general reader can well afford to miss."—*Pall Mall Gazette.*

"The last book published by Messrs. Merivale and Marzials is full of very real and true things."—Mrs. ANNE THACKERAY RITCHIE on "Thackeray and his Biographers," in *Illustrated London News.*

Life of Thoreau. By H. S. Salt.

"Mr. Salt's volume ought to do much towards widening the knowledge and appreciation in England of one of the most original men ever produced by the United States."—*Illustrated London News.*

Life of Voltaire. By Francis Espinasse.

"Up to date, accurate, impartial, and bright without any trace of affectation."—*Academy.*

Life of Whittier. By W. J. Linton.

"Mr. Linton is a sympathetic and yet judicious critic of Whittier."—*World.*

Complete Bibliography to each volume, by J. P. ANDERSON, British Museum, London.

New York: CHARLES SCRIBNER'S SONS.

The Contemporary Science Series.

Edited by Havelock Ellis.

12mo. Cloth. Price $1.50 per Volume.

I. THE EVOLUTION OF SEX. By Prof. PATRICK GEDDES and J. A. THOMSON. With 90 Illustrations. Second Edition.

"The authors have brought to the task—as indeed their names guarantee—a wealth of knowledge, a lucid and attractive method of treatment, and a rich vein of picturesque language."—*Nature.*

II. ELECTRICITY IN MODERN LIFE. By G. W. DE TUNZELMANN. With 88 Illustrations.

"A clearly written and connected sketch of what is known about electricity and magnetism, the more prominent modern applications, and the principles on which they are based."—*Saturday Review.*

III. THE ORIGIN OF THE ARYANS. By Dr. ISAAC TAYLOR. Illustrated. Second Edition.

"Canon Taylor is probably the most encyclopædic all-round scholar now living. His new volume on the *Origin of the Aryans* is a first-rate example of the excellent account to which he can turn his exceptionally wide and varied information. . . . Masterly and exhaustive."—*Pall Mall Gazette.*

IV. PHYSIOGNOMY AND EXPRESSION. By P. MANTEGAZZA. Illustrated.

"Brings this highly interesting subject even with the latest researches. . . . Professor Mantegazza is a writer full of life and spirit, and the natural attractiveness of his subject is not destroyed by his scientific handling of it."—*Literary World* (Boston).

V. EVOLUTION AND DISEASE. By J. B. SUTTON, F.R.C.S. With 135 Illustrations.

"The book is as interesting as a novel, without sacrifice of accuracy or system, and is calculated to give an appreciation of the fundamentals of pathology to the lay reader, while forming a useful collection of illustrations of disease for medical reference."—*Journal of Mental Science.*

VI. THE VILLAGE COMMUNITY. By G. L. GOMME. Illustrated.

"His book will probably remain for some time the best work of reference for facts bearing on those traces of the village community which have not been effaced by conquest, encroachment, and the heavy hand of Roman law."—*Scottish Leader.*

VII. THE CRIMINAL. By HAVELOCK ELLIS. Illustrated Fourth Edition, Revised and Enlarged.

"The sociologist, the philosopher, the philanthropist, the novelist—all, indeed, for whom the study of human nature has any attraction—will find Mr. Ellis full of interest and suggestiveness."—*Academy.*

New York: CHARLES SCRIBNER'S SONS.

XVII. THE HISTORY OF THE EUROPEAN FAUNA. By R. F. SCHARFF, B.Sc., Ph.D., F.Z.S. Illustrated.

XVIII. PROPERTY: ITS ORIGIN AND DEVELOPMENT. By CH. LETOURNEAU, General Secretary to the Anthropological Society, Paris, and Professor in the School of Anthropology, Paris.

"M. Letourneau has read a great deal, and he seems to us to have selected and interpreted his facts with considerable judgment and learning." —*Westminster Review.*

XIX. VOLCANOES, PAST AND PRESENT. By Prof. EDWARD HULL, LL.D., F.R.S.

"A very readable account of the phenomena of volcanoes and earthquakes."—*Nature.*

XX. PUBLIC HEALTH. By Dr. J. F. J. SYKES. With numerous Illustrations.

"Not by any means a mere compilation or a dry record of details and statistics, but it takes up essential points in evolution, environment, prophylaxis, and sanitation bearing upon the preservation of public health."—*Lancet.*

XXI. MODERN METEOROLOGY. AN ACCOUNT OF THE GROWTH AND PRESENT CONDITION OF SOME BRANCHES OF METEOROLOGICAL SCIENCE. By FRANK WALDO, Ph.D., Member of the German and Austrian Meteorological Societies, etc.; late Junior Professor, Signal Service, U.S.A. With 112 Illustrations.

"The present volume is the best on the subject for general use that we have seen."—*Daily Telegraph* (London).

XXII. THE GERM-PLASM: A THEORY OF HEREDITY. By AUGUST WEISMANN, Professor in the University of Freiburg-in-Breisgau. With 24 Illustrations. $2.50.

"There has been no work published since Darwin's own books which has so thoroughly handled the matter treated by him, or has done so much to place in order and clearness the immense complexity of the factors of heredity, or, lastly, has brought to light so many new facts and considerations bearing on the subject."—*British Medical Journal.*

XXIII. INDUSTRIES OF ANIMALS. By E. F. HOUSSAY. With numerous Illustrations.

"His accuracy is undoubted, yet his facts out-marvel all romance. These facts are here made use of as materials wherewith to form the mighty fabric of evolution."—*Manchester Guardian.*

New York: CHARLES SCRIBNER'S SONS.

XXIV. MAN AND WOMAN. By HAVELOCK ELLIS. Illustrated. Fourth and Revised Edition.

"Mr. Havelock Ellis belongs, in some measure, to the continental school of anthropologists; but while equally methodical in the collection of facts, he is far more cautious in the invention of theories, and he has the further distinction of being not only able to think, but able to write. His book is a sane and impartial consideration, from a psychological and anthropological point of view, of a subject which is certainly of primary interest."—*Athenæum.*

XXV. THE EVOLUTION OF MODERN CAPITALISM. By JOHN A. HOBSON, M.A. (New and Revised Edition.)

"Every page affords evidence of wide and minute study, a weighing of facts as conscientious as it is acute, a keen sense of the importance of certain points as to which economists of all schools have hitherto been confused and careless, and an impartiality generally so great as to give no indication of his [Mr. Hobson's] personal sympathies."—*Pall Mall Gazette.*

XXVI. APPARITIONS AND THOUGHT TRANSFERENCE. By FRANK PODMORE, M.A.

"A very sober and interesting little book. . . . That thought-transference is a real thing, though not perhaps a very common thing, he certainly shows."—*Spectator.*

XXVII. AN INTRODUCTION TO COMPARATIVE PSYCHOLOGY. By Professor C. LLOYD MORGAN. With Diagrams.

"A strong and complete exposition of Psychology, as it takes shape in a mind previously informed with biological science. . . . Well written, extremely entertaining, and intrinsically valuable."—*Saturday Review.*

XXVIII. THE ORIGINS OF INVENTION: A STUDY OF INDUSTRY AMONG PRIMITIVE PEOPLES. By OTIS T. MASON, Curator of the Department of Ethnology in the United States National Museum.

"A valuable history of the development of the inventive faculty."—*Nature.*

XXIX. THE GROWTH OF THE BRAIN: A STUDY OF THE NERVOUS SYSTEM IN RELATION TO EDUCATION. By HENRY HERBERT DONALDSON, Professor of Neurology in the University of Chicago.

"We can say with confidence that Professor Donaldson has executed his work with much care, judgment, and discrimination."—*The Lancet.*

XXX. EVOLUTION IN ART: AS ILLUSTRATED BY THE LIFE-HISTORIES OF DESIGNS. By Professor ALFRED C. HADDON. With 130 Illustrations.

"It is impossible to speak too highly of this most unassuming and invaluable book."—*Journal of Anthropological Institute.*

XXXI. THE PSYCHOLOGY OF THE EMOTIONS. By TH. RIBOT, Professor at the College of France, Editor of the *Revue Philosophique*.

"Professor Ribot's treatment is careful, modern, and adequate."—*Academy*.

XXXII. HALLUCINATIONS AND ILLUSIONS: A STUDY OF THE FALLACIES OF PERCEPTION. By EDMUND PARISH.

"This remarkable little volume."—*Daily News*.

XXXIII. THE NEW PSYCHOLOGY. By E. W. SCRIPTURE, Ph.D. (Leipzig). With 124 Illustrations.

XXXIV. SLEEP: ITS PHYSIOLOGY, PATHOLOGY, HYGIENE, AND PSYCHOLOGY. BY MARIE DE MANACÉÏNE (St. Petersburg). Illustrated.

XXXV. THE NATURAL HISTORY OF DIGESTION. By A. LOCKHART GILLESPIE, M.D., F.R.C.P. ED., F.R.S. ED. With a large number of Illustrations and Diagrams.

"Dr. Gillespie's work is one that has been greatly needed. No comprehensive collation of this kind exists in recent English Literature."—*American Journal of the Medical Sciences*.

XXXVI. DEGENERACY: ITS CAUSES, SIGNS, AND RESULTS. By Professor EUGENE S. TALBOT, M.D., Chicago. With Illustrations.

"The author is bold, original, and suggestive, and his work is a contribution of real and indeed great value, more so on the whole than anything that has yet appeared in this country."—*American Journal of Psychology*.

XXXVII. THE RACES OF MAN: A SKETCH OF ETHNOGRAPHY AND ANTHROPOLOGY. By J. DENIKER. With 178 Illustrations.

"Dr. Deniker has achieved a success which is well-nigh phenomenal."—*British Medical Journal*.

XXXVIII. THE PSYCHOLOGY OF RELIGION. AN EMPIRICAL STUDY OF THE GROWTH OF RELIGIOUS CONSCIOUSNESS. By EDWIN DILLER STARBUCK Ph.D., Assistant Professor of Education, Leland Stanford Junior University.

"No one interested in the study of religious life and experience can afford to neglect this volume."—*Morning Herald*.

XXXIX. THE CHILD: A STUDY IN THE EVOLUTION OF MAN. By Dr. ALEXANDER FRANCIS CHAMBERLAIN, M.A., Ph.D., Lecturer on Anthropology in Clark University, Worcester (Mass.). With Illustrations.

"The work contains much curious information, and should be studied by those who have to do with children."—*Sheffield Daily Telegraph*.

New York: CHARLES SCRIBNER'S SONS.

CPSIA information can be obtained
at www.ICGtesting.com
Printed in the USA
BVHW082058201118
533619BV00011B/1563/P